# Negation in Early English

Informed by detailed analysis of data from large-scale diachronic corpora, this book is a comprehensive account of changes to the expression of negation in English. Its methodological approach brings together up-to-date techniques from corpus linguistics and minimalist syntactic analysis to identify and characterise a series of interrelated changes affecting negation during the period 800–1700. Phillip W. Wallage uses cutting-edge statistical techniques and large-scale corpora to model changes in English negation over a period of nine hundred years. These models provide crucial empirical evidence which reveals the specific processes of syntactic and functional change affecting early English negation, and identifies diachronic relationships between these processes.

PHILLIP W. WALLAGE is Senior Lecturer in English Language and Linguistics at Northumbria University. His work on syntactic change in the history of English (principally focusing on negation) has been published in journals including *Lingua* and *English Language and Linguistics*.

STUDIES IN ENGLISH LANGUAGE



The aim of this series is to provide a framework for original studies of English, both present-day and past. All books are based securely on empirical research, and represent theoretical and descriptive contributions to our knowledge of national and international varieties of English, both written and spoken. The series covers a broad range of topics and approaches, including syntax, phonology, grammar, vocabulary, discourse, pragmatics and sociolinguistics, and is aimed at an international readership.

Already published in this series:

Christiane Meierkord: *Interactions across Englishes: Linguistic Choices in Local and International Contact Situations*
Haruko Momma: *From Philology to English Studies: Language and Culture in the Nineteenth Century*
Raymond Hickey (ed.): *Standards of English: Codified Varieties Around the World*
Benedikt Szmrecsanyi: *Grammatical Variation in British English Dialects: A Study in Corpus-Based Dialectometry*
Daniel Schreier and Marianne Hundt (eds.): *English as a Contact Language*
Bas Aarts, Joanne Close, Geoffrey Leech and Sean Wallis (eds.): *The Verb Phrase in English: Investigating Recent Language Change with Corpora*
Martin Hilpert: *Constructional Change in English: Developments in allomorphy, word formation, and syntax*
Jakob R. E. Leimgruber: *Singapore English: Structure, Variation and Usage*
Christoph Rühlemann: *Narrative in English Conversation*
Dagmar Deuber: *English in the Caribbean: Variation, Style and Standards in Jamaica and Trinidad*
Eva Berlage: *Noun Phrase Complexity in English*
Nicole Dehé: *Parentheticals in Spoken English: The Syntax-Prosody Relation*
Jock Onn Wong: *English in Singapore: A Cultural Analysis*
Anita Auer, Daniel Schreier and Richard J. Watts: *Letter Writing and Language Change*
Marianne Hundt: *Late Modern English Syntax*

Irma Taavitsainen, Merja Kyto, Claudia Claridge, and Jeremy Smith: *Developments in English: Expanding Electronic Evidence*

Arne Lohmann: *English Co-ordinate Constructions: A Processing Perspective on Constituent Order*

John Flowerdew and Richard W. Forest: *Signalling Nouns in English: A corpus-based discourse approach*

Jeffrey P. Williams, Edgar W. Schneider, Peter Trudgill, and Daniel Schreier: *Further Studies in the Lesser-Known Varieties of English*

Nuria Yáñez-Bouza: *Grammar, Rhetoric and Usage in English: Preposition Placement 1500–1900*

Jack Grieve: *Regional Variation in Written American English*

Douglas Biber and Bethany Gray: *Grammatical Complexity in Academic English: Linguistics Change in Writing*

Gjertrud Flermoen Stenbrenden: *Long-Vowel Shifts in English, c. 1050–1700: Evidence from Spelling*

Zoya G. Proshina and Anna A. Eddy: *Russian English: History, Functions, and Features*

Raymond Hickey: *Listening to the Past: Audio Records of Accents of English*

Earlier titles not listed are also available

# Negation in Early English

## Grammatical and Functional Change

PHILLIP W. WALLAGE

*Northumbria University, Newcastle*

CAMBRIDGE
UNIVERSITY PRESS

**CAMBRIDGE**
**UNIVERSITY PRESS**

University Printing House, Cambridge CB2 8BS, United Kingdom

One Liberty Plaza, 20th Floor, New York, NY 10006, USA

477 Williamstown Road, Port Melbourne, VIC 3207, Australia

4843/24, 2nd Floor, Ansari Road, Daryaganj, Delhi – 110002, India

79 Anson Road, #06–04/06, Singapore 079906

Cambridge University Press is part of the University of Cambridge.

It furthers the University's mission by disseminating knowledge in the pursuit of education, learning, and research at the highest international levels of excellence.

www.cambridge.org
Information on this title: www.cambridge.org/9781107114296
DOI: 10.1017/9781316335185



First published 2017

*A catalogue record for this publication is available from the British Library*

*Library of Congress Cataloging-in-Publication Data*
NAMES: Wallage, Phillip, 1978–
TITLE: Negation in early English : grammatical and functional change / Phillip Wallage.
DESCRIPTION: Cambridge ; New York : Cambridge University Press, [2017] |
Series: Studies in English language | Includes bibliographical references and index.
IDENTIFIERS: LCCN 2016059627 | ISBN 9781107114296 (hard back : alk. paper)
SUBJECTS: LCSH: English language–Negatives. | English language–Old English, ca. 450-1100–Grammar. | English language–Middle English, 1100-1500–Grammar. | English language–Variation. | Linguistic change. | Historical grammar.
CLASSIFICATION: LCC PE1359 .N44 [2017] | DDC 425–dc23 LC record available at https://lccn.loc.gov/2016059627

ISBN 978-1-107-11429-6 Hardback

For Mum and Dad

# Contents

# Figures

# Tables

# Acknowledgements

This book represents the culmination of over a decade of research on negation that began with my doctoral studies. During that time, I received advice, suggestions, support and encouragement from many people. All have contributed to making this book better than it otherwise would be.

I would like to acknowledge the financial support of the Arts and Humanities Research Council (UK), who funded my doctoral studies; the Economic and Social Research Council (UK) who awarded me a postdoctoral fellowship; The University of Manchester, where I held a Research Fellowship; and Northumbria University, for several periods of research leave.

I owe a debt of gratitude to Anthony Warner, who supervised my doctoral work and got me started on this topic in the first place. During that time, Anthony gave me lots of what has proved to be excellent advice. Others whose advice and support have proved invaluable are Susan Pintzuk, Eric Haeberli, Ann Taylor, Wim van der Wurff, David Denison, Richard Hogg and Ewa Dabrowska. I'm grateful for the support of my friends and colleagues at the Universities of York, Manchester and Northumbria during the times when negativity threatened to overwhelm me, in particular Beck Sinar, Marion Elenbaas and Nuria Yanez-Bouza. I appreciate the patience of my students, who have heard more about negation – and about the process of writing this book – than I'm sure they ever wanted to. Thanks too to Andrew Winnard, Mary Catherine Bongiovi and the team at Cambridge University Press, who have been very accommodating with their deadlines, patient with my queries and have made the process of publishing very straightforward.

Earlier versions of some of this research have been presented at numerous conferences and published as Wallage (2008, 2012a, 2013). Over the last decade and more, I've benefitted from discussions with and comments from far too many people to list here, but including David Adger, Lauren Alsenoy, Johan van der Auwera, Montserrat Batllori, Linda van Bergen, Theresa Biberauer, Anne Breitbarth, Aaron Ecay, Elly van Gelderen, Lilane Haegeman, Maj-Britt Moesgaard Hansen, Glyn Hicks, Anders Holmberg, Laurence Horn, Richard Ingham, Ans van Kemenade, Anthony Kroch, Pierre Larrivee, Roger Lass, Chungmin Lee, Chris Lucas, Paul Rowlett,

Graeme Trousdale, Jacqueline Visconti, George Walkden, Joel Wallenburg, David Willis and Hedde Zeijlstra.

My most heartfelt thanks goes to my parents, who have been the greatest source of support and encouragement as I wrote the book, and throughout my life. Without their unfaltering belief in me, I would not have been able to do any of this. This book is dedicated to them.

# Abbreviations

**BNC** British National Corpus
**C.** Century
**CE** Common Era or Current Era
**C-I** Conceptual-intensional interface
**COCA** Corpus of Contemporary American English
**EME** Early Middle English, c.1150–1350CE
**iNeg** LF-interpretable negative feature
**JC** Jespersen Cycle
**LF** Logical Form
**LME** Late Middle English, c.1350–1500CE
**ME** Middle English, c.1150–1500CE
**NC** Negative concord
**ND** Negative doubling
**NegV1** Clause with a negated finite verb in first position
**NI** Negative item
**NM** Negative marker
**NPI** Negative polarity item
**n-word** negative word
**OE** Old English, c.800–1100CE
**Op** Operator
**PCEEC** *Parsed Corpus of Early English Correspondence* (Taylor et al., 2006)
**PDE** Present-day English
**PF** Phonological Form
**PN** Paratactic or redundant negation
**PPCEME** *Penn-Helsinki Parsed Corpus of Early Modern English* (Kroch et al., 2004)
**PPCME2** *Penn-Helsinki Parsed Corpus of Middle English* (2nd edition) (Kroch and Taylor, 2000)
**QC** Quantifier Cycle
**uNeg** LF-uninterpretable negative feature
**YCOE** *York-Toronto-Helsinki Parsed Corpus of Old English Prose* (Taylor et al., 2002)

# 1 Introduction

## 1.1 The Jespersen Cycle and Theories of Syntactic Change

### *1.1.1 The Jespersen Cycle*

The Jespersen Cycle (Jespersen, 1917) has long been regarded as a textbook case of grammaticalisation, found in many languages. The English Jespersen Cycle manifests itself as a sequence of overlapping stages, each formally distinct from the preceding one as in (1)–(3).

(1) Stage One: Sentential negation is marked by *ne* alone (c.1150–1300)

a. we **ne** moten halden Moses e lichamlice
we NEG might observe Moses' law bodily
'we might not observe Moses law literally'
(CMLAMBX1,89.735)

b. we **ne** mugen þat don
we NEG can that do
'We cannot do that'
(CMTRINIT,103.1370)

(2) Stage Two: The sentential negator *not* co-occurs with *ne*. Sentential negation comprises two parts (c. 1150–1400)

a. ac of hem **ne** speke ic **noht**
but of them NEG spoke I not
'but I did not speak of them'
(CMTRINIT,95.1272)

b. I **ne** may **nat** denye it
I NEG may not deny it
'I may not deny it'
(CMBOETH,435.C1.262)

(3) Stage Three: Sentential negation is marked by *not* alone. (c.1350–1500)

a. Thou shalt **not** do so
You ought not do so
'You ought not do so'
(CMROLLTR,41.855)

b. I know **nat** the cause
I know not the cause
'I do not know the cause'
(CMMALORY,627.3549)

The changes result in the grammaticalisation of *not* as a marker of sentential scope negation, and take place during a period from the ninth century to the fifteenth century. What changes constitute the Jespersen Cycle?

Much has been written about the process by which new negative markers are grammaticalised – there are many syntactic, semantic and pragmatic accounts of the processes involved in many languages, each of which decompose the change into different stages, and hence conceptualise the mechanisms and diachrony of the Jespersen Cycle in different ways. Existing theoretical accounts differ fundamentally in their conceptions of the cycle – how many stages are involved, what the syntactic elements and configurations are at each stage, what formal linguistic mechanisms are invoked to explain the change, whether the cycle involves morphological, syntactic or functional change at each stage and how these various factors might interact.

Given that formal and functional accounts structure the cycle in different ways and invoke different mechanisms of change, what evidence do we have to decide between them? Each account makes different predictions about how the three stages *ne, ne...not* and *not* relate to each other, what changes are involved in the cycle and how those changes progress over time. An account of the Jespersen Cycle must not only formalise each diachronic stage, but also allow us to model the changes in a way that predicts the distribution of *ne, ne...not* and *not* observed in diachronic data as the changes progress. Formal and functional linguistic analyses should also structure the Jespersen Cycle in ways that are consistent with what we know about how morphosyntactic changes progress over time.

In this book, I argue that – viewed in this way – quantitative models based on data from diachronic corpora provide crucial empirical evidence to inform formal accounts of the Jespersen Cycle, and to establish what mechanisms of change are involved. By placing strict empirical constraints on which changes are plausible and which changes are not, quantitative models of change provide a solid foundation upon which to build formal and functional analyses of the Jespersen Cycle, and to ascertain how different formal and functional changes relate to each other within the cycle.

This empirical approach provides new evidence that *ne* undergoes morphosyntactic weakening prior to its loss. This idea is first proposed by Jespersen (1917), but has proven problematic to formalise, and is often not captured in syntactic analyses of the Jespersen Cycle such as Haegeman (1995), Roberts and Roussou (2003) and Zeijlstra (2004). Corpus evidence enables us to formalise the steps in this weakening process very precisely and to identify their empirical effects. The morphosyntactic weakening of *ne* is not only crucial to the concomitant grammaticalisation of *not*, its effects go far beyond the Jespersen Cycle. When properly formalised, the weakening of *ne* provides a coherent explanation of several interlinked changes to negation during the Old English (c.800–1100) and Middle English (c.1100–1500) periods. These include changing patterns of redundant negation and negative inversion, changes in the form and availability of negative concord and the relationship between the loss of negative concord and the Jespersen Cycle.

### *1.1.2 Morphosyntactic Change and the Jespersen Cycle*

How can we analyse the Jespersen Cycle in a way that is consistent with formal approaches to morphosyntactic variation and change? The notion of syntactic parameter is central to generative models of syntactic variability, in order to explain relationships between languages, varieties or diachronic stages within the same language variety; to constrain possible axes of variation across languages, varieties or diachronic stages; and to establish pathways of morphosyntactic change.

Typically, in parameter-based approaches to morphosyntactic change, change proceeds through competition between two formally distinct but functionally equivalent syntactic options. These options constitute different ways to realise or set a particular parameter. For example Pintzuk (1999) argues that Old English VPs may either be head-initial or head-final. In sixteenth century English, Kroch (1989) argues that the tense head may be lexicalised by V to T movement or by insertion of DO in $T^0$. Variation or change in a single parameter may explain the presence or absence of several surface syntactic phenomena that are the reflexes of that parameter. For example Bobaljik and Thrainsson (1998) argue that whether or not a language projects Agreement (Agr) projections has several consequences including the availability of distinct markers for tense and agreement on finite verbs, object shift and transitive expletive constructions. Kroch (1989) proposes that all reflexes of a particular parametric shift should change at a constant rate over time, providing an empirical basis to identify which changes are manifestations or reflexes of a single parameter.

However, an approach to grammatical competition in terms of binary parameters is problematic for the Jespersen Cycle because, at least on the surface, it appears too restrictive. The Jespersen Cycle does not involve straightforward substitution of one form (*ne*) for another (*not*). Stage two of the Jespersen Cycle is particularly problematic. The Middle English Jespersen Cycle comprises at least three stages (1)–(3) given above, with apparent redundancy in negative marking in the second stage (the co-occurrence of *ne* and *not*). Deriving three stages requires a sequence of at least two parametric changes, raising the question of what these two parametric changes are and how they relate to each other. Notions of functional equivalence and mutual exclusivity pose difficulties too. Clearly, *ne* and *not* are not mutually exclusive at stage two, therefore they are not functionally equivalent. Many accounts of the Jespersen Cycle such as Frisch (1997), Roberts and Roussou (2003) and Zeijlstra (2004) regard the grammaticalisation of *not* and the loss of *ne* as independent but intersecting changes, each involving a different parameter. Under this approach, the co-occurrence of *ne* and *not* at stage two is epiphenomenal. However, we will see that *ne...not* is distributionally independent of the preceding and following stages. Modelling the distribution of *ne...not* therefore requires a fundamentally different account of the parameters involved.

The status and place of parameters has been reappraised in recent formal (Minimalist) syntactic frameworks. Recent Minimalist approaches such as Chomsky (1999, 2001), and accounts within this paradigm that seek to account for variation, such as Roberts and Roussou (2003), Adger (2006) and Adger and Trousdale (2007), locate parametric variation in the lexicon, operationalising Borer's (1984) insight. By making the morphosyntactic features associated with lexical items the triggers for certain syntactic operations like agreement or movement, these accounts place very tight constraints on what constitutes a parameter and what values or settings a parameter may have. This constrains our approach to morphosyntactic competition. Competition is not between entire grammars or grammatical subsystems, but a choice between lexical items with mutually exclusive feature specifications all present in the lexicon of an individual speaker. I demonstrate that the Jespersen Cycle can be analysed as a sequence of parametric changes in these terms, and furthermore that such an account receives empirical support from patterns of variation and change in diachronic corpora. It also has the conventional advantage associated with parametric accounts of variation – the ability to explain several changes in the grammar of negation as reflexes of a single parametric shift.

## 1.2 The Jespersen Cycle and Other Aspects of the Grammar of Negation

Several negation phenomena in early English undergo change. I argue that these changes are not only linked to the parametric changes underlying the Jespersen Cycle, but that examination of these changes informs an account of the cycle. They include changes to redundant negation, negative inversion and negative concord.

Of these, the relationship between the Jespersen Cycle and negative concord is the most widely discussed (see for example Haegeman (1995); Haegeman and Zanuttini (1996); Rowlett (1998); Zeijlstra (2004, 2008)). Languages are either typically[1] negative concord languages or double

[1] See Larrivée (2015) for discussion of double negation readings in negative concord languages.

negation languages. In negative concord, as the Middle English example in (4) illustrates, a clause contains more than one negative word but it receives a reading of sentential scope negation irrespective of how many negative words it contains.

(4) but he was so hard, þat **no** begger might gete **no** good of hym by **no**
but he was so hard that no beggar might get no good of him in no
maner wyse
manner way
'But he was so hard-hearted that no beggar might get any good of him in any kind of way'
(15th century; MIRK,104.2825)

Contrast that with standard Present-day English, in which each negative form contributes negation to the clause. Thus, Present-day standard English (5) with two negative forms receives an affirmative reading – each negative form is interpreted as negative so their effect is to cancel each other out.

(5) No-one said nothing at the meeting = Everyone said something

Many non-standard varieties of Present-day English (PDE) exhibit negative concord but Standard English does not. Nevalainen (1996) argues that this change is, in part, linked to the standardisation of English in the sixteenth century. However, as I argue in Chapters 7–9, this does not entirely explain the decline in negative concord, which begins much earlier in the thirteenth and fourteenth centuries. I argue that this earlier decline in negative concord is a consequence of the same parametric changes that drive the Jespersen Cycle. Furthermore, I argue that the distinction between not-negation (6a) and no-negation (6b) observed by Tottie (1991a) and the distribution of these two variants in both historical and Present-day English falls out of this account.

(6) a. I didn't see anyone
b. I saw no-one

The other two phenomena I focus on – redundant negation and negative inversion – are less well studied. A redundant negative is a negative word that appears in the complement clause of certain verbs such as *deny, forbid or prohibit*, but does not contribute negative force to the clause, as illustrated by (7).

(7) You may deny that you were **not** the meane of my Lord Hastings late
You may deny that you were not the cause of my Lord Hastings late
imprisonment
imprisonment

'You may deny you were the cause of my Lord Hastings's recent imprisonment'
(Richard 3, I.iii.502–503, van der Wurff 1999, 301, ex.14)

Present-day English negative inversion is described by Klima (1964) and is central to the analyses of negation proposed by Haegeman (1995, 2001). The term describes the pattern seen in example (8).

(8) Never have I been so reviled
(BNC, Margret Forster Lady's Maid, c.1990)

When a clause-initial negative takes sentential scope, we find subject-operator inversion. The canonical order of subject–verb found in English declaratives is reversed. Van Kemenade (2001) identifies a similar pattern in Old English as early as the ninth century.

The distributions of both redundant negation and negative inversion change over time. Their availability and forms differ at successive stages of the Jespersen Cycle in ways that are problematic to previous accounts of redundant negation such as van der Wurff (1999b), or negative inversion such as Nevalainen (1997) and Ingham (2007), but in ways that inform the analysis of each stage of the Jespersen Cycle in the present account.

## 1.3 Methodological Approach

### *1.3.1 Modelling Morphosyntactic Change Using Diachronic Corpora*

Previous accounts of the Jespersen Cycle have sought their evidence in historical data, but taken a largely qualitative view of those data. By contrast, the present analysis develops a mixed methodological approach to corpus data which is not commonly practiced in analyses of diachronic change.

On an empirical level, any theoretical account should be testable against naturally occurring linguistic data, such as data from corpora of historical texts. However, there are various ways of using corpora as evidence in historical linguistics. The most common is as a source of qualitative data. Thus the existence of certain forms or patterns in the data at certain periods of time are the phenomena that a theoretical analysis must account for. This is a synchronic approach to data from successive periods of time. Such accounts are focused on the inputs to and products of change. Processes of change are inferred from qualitative data. This kind of analysis will inform the analysis of *ne* and *not* in Chapters 3–5 and the analysis of negative concord in Chapter 8. However, I will argue it alone is not sufficient to characterise the Jespersen Cycle.

A growing body of recent work in historical linguistics has become informed by the quantitative methods involved in variationist sociolinguistics. This work uses corpora not only as a source of synchronic data, but also as a basis from which to model diachronic change in progress. For example, see the work of Kroch and his colleagues on morphosyntactic change, beginning with Kroch (1989); or the work of Nevalainen and her colleagues on historical sociolinguistics, for example Nevalainen and Ramoulin-Brunberg (2003). These accounts focus on diachronic change as a process and aim to account for the distribution of forms undergoing change.

Change proceeds through competition between mutually exclusive and functionally equivalent forms which represent competing ways to set a syntactic parameter. Morphosyntactic change proceeds along a logistic curve, therefore we can model changes in progress using logistic regression and compare constraints on the distribution of competing forms at different points in time. Each formal account of the stages involved in the Jespersen Cycle implicitly or explicitly provides a basis for modelling the diachrony of the Jespersen Cycle: how we structure the parametric shifts within the Jespersen Cycle makes predictions about the distribution of competing forms as change progresses in corpus data. Kroch (1989, 235, fn.29) suggests:

> Once the principle that contexts change together when they are surface reflexes of a single grammatical competition becomes firmly established, it may be possible, on occasion, to choose among grammars proposed on the basis of synchronic analysis by the predictions they make as to which contexts should change together.

Syntactically independent forms have independent distributions in corpus data, whereas the distributions of forms that are reflexes of a single syntactic parameter will pattern together in diachronic corpus data. Changes that are reflexes of a single parameter should observe the Constant Rate Effect throughout the course of the change. As syntactic analyses structure the parameters involved in the Jespersen Cycle in different ways, so they make different predictions about what competes with what, how variation and change in the distribution of forms is structured, and how it will pattern over time. These predictions are tested against observed patterns of variation and change in corpus data using logistic regression models. This places empirical constraints on possible analyses: only analyses which accurately predict the distributions of *ne* and *not* we observe in the diachronic corpora can be considered appropriate analyses of the Jespersen Cycle.

For example, if the Jespersen Cycle involves a single form of *ne*, *ne* should be subject to the same distributional constraints throughout the Jespersen Cycle, and these constraints should be constant over time. On the other hand, if as I hypothesise, the Jespersen Cycle involves syntactically different forms of *ne* at stage one and stage two, then their distributions in the corpus data will be independent. This provides a very precise empirical basis to evaluate different formal hypotheses of change.

### 1.3.2 Modelling Functional Change

In Chapter 6, I argue that this methodological approach also provides an empirical basis to ascertain the role of functional change within the Jespersen Cycle. Parametric accounts of change face the issue of redundancy. Why would a language have two ways of expressing the same thing, particularly if those ways are functionally equivalent and differ only in form? In one sense, locating parametric variation in the lexicon lessens this problem – lexical synonyms provide some evidence for redundancy. However synonyms may exhibit subtle differences in meaning or differ in their pragmatics. By extension, we might ask on what linguistic levels competing grammatical forms are equivalent and on what levels they are distinct.

Recent accounts of the Jespersen Cycle in languages other than English, principally for French (Detges and Waltereit, 2002; van der Auwera, 2009; Hansen, 2009; Hansen and Visconti, 2009), make functional change fundamental to the grammaticalisation of new negative markers. New negative markers emerge in pragmatically marked contexts, and gradually come into competition with the established negative marker as they spread from pragmatically marked to pragmatically unmarked environments. These accounts imply that functional constraints on new forms weaken as their frequency increases. This challenges Kroch's observation that the the constraints on an innovative form, when estimated probabilistically, will remain constant over the course of the change. Thus accounts invoking functional change make predictions about the diachrony of the Jespersen Cycle, which can be modelled statistically and tested for fit against diachronic corpus data. By testing the distributions of *ne, ne...not* and *not* predicted by these statistical models against the observed distributions of *ne, ne...not* and *not* in corpora, not only can we establish the number of stages required in a model of the Jespersen Cycle and their appropriate syntactic analysis, we can also identify functional shifts involved in the cycle, and model the interaction between the syntactic changes and these functional changes.

### 1.3.3 Diachronic Corpora

Given the mixed methodological approach I adopt, corpora are crucial to this work, not only as sources of qualitative data, but also as sources of quantitative data. The way the distribution of forms patterns over time can only be examined using large scale diachronic corpora that cover a long timespan. Thus the use of such corpora is essential to test the fit of models of morphosyntactic and functional change.

For such work, the corpora must meet a number of criteria. As in any corpus study, the corpus must be a representative sample of the language varieties under examination. However, particularly in studies of change, it is important that the corpus design is such that when the data are subdivided into periods for diachronic analysis, the data from each of those successive sub-periods are comparable. This ensures that changes in the frequency of a form across successive sub-periods is in fact diachronic change rather than variation resulting from inconsistencies in the composition of the corpus from one period to the next. This problem is exacerbated in historical linguistic research because of the paucity of sources available to us and the narrow demographic of individuals who produced written texts. Therefore there are concerns about whose language they can be said to represent. Despite this, the corpora must be balanced as far as possible from one sub-period to the next. Syntactically parsed corpora are preferred for the investigation of morphosyntactic variation and change, simply because syntactic structures are the object of study and such corpora make those structures easier to interrogate. Finally, as much of the analysis is concerned with testing the fit of models of diachronic change to corpora, in quantitative and statistical terms, the corpora need to be large enough to allow statistical analysis to be performed. On this basis, the corpora selected are the Penn and York historical corpora as follows:

| Date | Corpus | Word-count |
|---|---|---|
| 800–1100 | York Helsinki Parsed Corpus of Old English Prose (YCOE) (Taylor et al., 2002) | 1.5 million |
| 1100–1500 | Penn Helsinki Parsed Corpus of Middle English 2nd edition (PPCME2) (Kroch and Taylor, 2000) | 1.2 million |
| 1500–1700 | Parsed Corpus of Early English Correspondence (PCEEC, Taylor et al. (2006)) | 2.2 million |
| 1500–1700 | Penn Helsinki Parsed Corpus of Early Modern English (PPCEME) (Kroch et al., 2004) | 1.7 million |
| | | 6.6 million |

For Present-day English, the British National Corpus (BNC) and the Corpus of Contemporary American English (COCA) are used. The BNC is a 100 million word corpus of spoken and written British English collected and compiled during the 1980s and early 1990s. COCA is a 520 million word corpus of spoken and written American English dating from 1990 to 2015. Both are grammatically tagged, but not syntactically parsed.

## 1.4 Outline of the Study

I hypothesise that bipartite *ne...not* is split into two syntactically distinct and competing *ne...not* stages. In the first of these, *ne* is the negative marker, and in the second the negative marker is *not*. The bipartite *ne...not* construction involves (at least) two syntactically and functionally distinct competing forms of *not* and two syntactically and functionally distinct competing forms of *ne*. The analysis therefore needs to establish where *ne* and *not* are negative markers and then what *ne* and *not* are when they are not negative markers.

I argue that formalising the syntactic differences between the two types of *ne* and the two types of *not* requires a morphosyntactic feature based account, making use of the distinction between semantically interpretable and semantically uninterpretable formal features invoked in Minimalist frameworks. Competition between semantically interpretable and semantically uninterpretable negative items derives changing patterns of redundant negation, negative inversion and – in combination with a quantifier cycle – the changing availability and form of negative concord in Middle English.

The argument proceeds as follows, beginning with the Jespersen Cycle. Chapter 2 presents evidence from quantitative data that demonstrate there are two competing forms of *ne*. Chapter 3 presents evidence to characterise each form of *ne*. Chapter 4 presents evidence for two distinct forms of *not* and evidence to characterise them syntactically and functionally. Chapter 5 then presents a formal analysis of the Jespersen Cycle, focusing on how the relationship between *ne* and *not* at successive stages is formalised in terms of features and dependencies. Chapter 6 examines functional change within the Jespersen Cycle. The patterning of functional constraints over time provides empirical evidence to identify the place of functional change within the cycle and argue for a particular relationship between formal and functional change.

Chapters 7 and 8 turn to negative concord, identifying and formalising links between the Jespersen Cycle, changes to negative concord, and the availability of negative inversion. Chapter 9 extends this analysis, arguing that variation between no-negation and not-negation is a reflex of the Jespersen Cycle that persists into Present-day English. Chapter 10 concludes the study with a model of the Jespersen Cycle, and discusses its implications for theories of morphosyntactic change.

# 2 Quantitative Evidence for a Model of the Jespersen Cycle in Middle English

## 2.1 Introduction

This chapter discusses different syntactic approaches to the Jespersen Cycle. These syntactic analyses structure the cycle differently, postulating different pathways of diachronic change that make different predictions about the distribution of negative markers during the Middle English period. These predictions form the basis for quantitative models of the Jespersen Cycle that are tested for goodness of fit against the actual distribution of negative markers in the PPCME2 corpus (Kroch and Taylor, 2000).[1] The results indicate that only one of these syntactic approaches can model the distribution of negative markers in diachronic corpus data in a way that both captures the observed distributions of *ne, ne...not* and *not*, and is consistent with the principles of morphosyntactic competition outlined by Kroch (1989).

Analyses of the Jespersen Cycle which assume a generative approach to syntactic theory fall into two broad groups with respect to how they structure the Jespersen Cycle. One group, including Haegeman (1995), Frisch (1997), Roberts and Roussou (2003) and Zeijlstra (2004), hypothesise that the early stages of the Jespersen Cycle represent only a change in the forms associated with particular syntactic positions. At both stage one (*ne*) and stage two (*ne... not*), they assume the same underlying syntactic structures and configurations – a NegP, in which there is agreement between a specifier (spec,NegP) and a head ($Neg^0$). At stage one, spec,NegP hosts a null negative operator. At stage two, the negative operator is overt, lexically realised by *not*. Thus the transition from *ne* to *ne... not* is a change in the way the spec,NegP position is realised – a change in the form of the negative operator.

The development of the negative operator *not* proceeds through grammaticalisation of a negative adverb. The etymological origin of *not* is Old English *nawiht* 'no creature'. It undergoes a category shift from nominal to adverbial and comes to be used as a minimiser in certain pragmatic contexts

[1] The goodness of fit can be estimated by examining whether the distributions of *ne, ne...not* and *not* predicted in each variable context under each model differ significantly ($p \leq .05$) from the distributions observed in corpus data.

(see Chapter 4 for a more detailed history of *not*). Thus accounts of the Jespersen Cycle such as Frisch (1997) postulate an intermediate stage in the development of bipartite forms, in which *not* is not a negative operator but a negative adverb or minimiser. The syntactic configurations involved are given in (9).

(9) a. Stage one: null Op[NEG] … *ne*
b. Stage two: null Op[NEG] … *ne* … *nawiht*
c. Stage three: *not*[NEG] … *ne*
d. Stage four: *not*[NEG]

As set out in (9), the Jespersen Cycle involves the grammaticalisation of *not* and the loss of *ne*, presumably through a process of phonetic erosion, as a weakly stressed proclitic on the finite verb, although this process is not described in detail in Frisch (1997) or the other accounts of this kind. In this model, bipartite *ne...not* is structurally ambiguous. The main difference between stages two and three is in the form of the negative operator. At stage two, *not* is a concordant negative adverb licensed by a null negative operator. At stage three, it is a negative operator. The question arises of how these two stages in the development of *not* can be identified and distinguished in corpus data. I discuss this issue in Section 2.2.1 and again in Chapter 4. More important for the present discussion is how these approaches conceptualise the changes involved in the Jespersen Cycle. The grammaticalisation of the negative adverb *nawiht* as *not* is a morphological change in the way the negative operator is realised. These models involve two syntactically distinct forms of *not* (an adverb and an operator), but only one form of *ne* whose syntax remains uniform throughout the Jespersen Cycle.

The second group of analyses includes Rowlett (1998), van Gelderen (2008), Breitbarth (2009), Willis (2011a) and my own earlier work (Wallage, 2005, 2008, 2012b). Although the details of each approach differ in ways that will become important in Chapter 5, they share in common one key point – they all argue that the underlying syntactic configurations of negative elements change during the Jespersen Cycle, rather than just the forms of those elements. For these analyses, the transition from simple negation (i.e. English *ne*) to bipartite negation (i.e. English *ne… not*) is not simply a change in the morphological form of an underlying negative operator in a specifier of NegP. Instead, it represents a shift in the syntactic and semantic locus of negation in negative clauses from *ne* to *not*. Two syntactically distinct forms of *ne* are required in order to represent this change: one in which *ne* is a negative marker, the other in which it is not.

When *ne* is the sole negative element in the clause, it is a negative marker contributing a negative interpretation to the clause at logical form (LF). However, once *not* becomes grammaticalised, *not* is the element which contributes a negative interpretation to the clause at LF. Of course, this raises the question of what the function of *ne* is in the *ne… not* construction. On this point, the analyses differ. I return to a detailed discussion of this issue in Chapter 3. For the purposes of the current discussion, it is sufficient to observe that there are two forms of *ne*: one which can stand alone in the clause and function as a negative marker, the other which cannot stand alone and therefore does not function as a negative marker. This group of approaches broadly schematise the Jespersen Cycle as in (10). These accounts make the assumption that negation is not associated with a single position at LF, but can be marked in different positions, providing it has sentential scope. The shift from stage two to stage three in the schema (10) represents a shift in the locus of negation from *ne* ($Neg^0$) to *not* (spec,NegP).

(10) a. Stage one: *ne*[NEG]
b. Stage two: *ne*[NEG] … *nawiht* (concordant negative adverb)
c. Stage three: *ne* … *not*[NEG] (negative marker)
d. Stage four: *not*[NEG] (negative marker)

There is one crucial difference between this schema and the schema in (9). In (10), the grammaticalisation of *not* as a negative marker is accompanied by a concomitant change in the syntax and semantics of *ne*, which loses its ability to mark negation as *not* becomes grammaticalised, whereas in (9), the grammaticalisation of *not* does not affect the syntax of *ne* – it is a concordant negative item rather than a negative marker throughout the Jespersen Cycle. So the schema in (9) requires two forms of *not* (adverb and negative marker) but only one form of *ne*, while the schema in (10) requires both two forms of *not* and two forms of *ne*.

In this chapter, I show that the models in (9) and (10) make different quantitative estimates of the frequency and distribution of *ne* across different grammatical contexts, which can then be tested for fit against the distribution of forms in corpus data. It follows from the model in (9) that *ne* undergoes the same change at stages one, two and three of the Jespersen Cycle, and that its distribution will be the same at all three stages. The loss of *ne* should proceed at the same rate irrespective of whether it appears alone at stage one or co-occurs with *not* at stages two or three. Conversely, different distributions of *ne* at stages one and two versus at stage three of the Jespersen Cycle provide evidence for the schema in (10). These differences may be qualitative, as I argue in Chapters 3 and 4 – that *ne* appears in different syntactic environments or configurations at stages one and two versus stage three of the cycle. However, differences between *ne* at stages one, two and three of the cycle may manifest themselves quantitatively. The two types of *ne* will be diachronically independent, lost at different rates in different processes of competition with different grammatical competitors, and their distributions subject to different contextual constraints.

## 2.2 Analyses That Assume One Form of *ne*

### 2.2.1 *Frisch's (1997) 'Redundant Licensing' Model of the Jespersen Cycle*

Frisch (1997) assumes that negative clauses have a NegP which comprises a specifier position and a head position as in (11). The structure of NegP is uniform throughout the Jespersen Cycle.

(11)

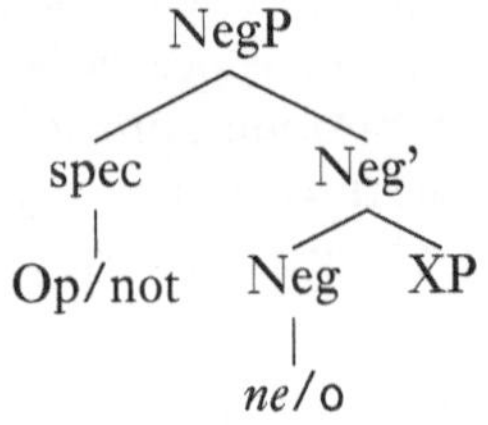


He argues that negation must be morphologically identified by an overt negative word. Either position within NegP can be lexically overt. When both positions are overt, NegP is redundantly licensed – identified by two overt negative forms. Under this model, there are two independent processes of competition – $Neg^0$ is either overt *ne* or null and spec,NegP hosts either a null operator or the operator *not*. Bipartite *ne* … *not* is an epiphenomenon, resulting from the intersection of these two independent changes in $Neg^0$ and spec,NegP. The result is a very precise quantitative prediction, given in (12). The frequency of *ne*… *not* results from the intersection of the overall frequencies of *not* in the spec,NegP position and of *ne* in $Neg^0$.

(12) Probability of *ne...not* = Probability *ne* is used x Probability *not* is used (adapted from Frisch 1997: 51)

Table 2.1 compares the frequencies of *ne*… *not* predicted by (12), with the frequencies of *ne...not* attested in diachronic data from the PPCME2 corpus.[2] The PPCME2 data are divided into four sub-periods. The data in Table 2.1 include instances of sentential scope negation involving *ne, ne...not* or *not* in finite main and subordinate clauses. Clauses involving negative indefinite determiners of NPs (as in *no beast* (13a)), negative indefinite pronouns (as in *noman* (13b)) or negative adverbials (such as *never* (13c)) are excluded from the figures. As Frisch (1997, 57) observes, these pattern very differently with regard to the Jespersen Cycle. See Chapters 7–9 for detailed discussion of clauses with negative indefinites.

(13) a. And þerefore may **no** best lyue þere
And therefore may no beast live there
'And therefore no animal may live there'
(CMMANDEV,10.199)

[2] All prose texts from the PPCME2 are included (for further details of the composition of the corpus see www.ling.upenn.edu/hist-corpora/PPCME2-RELEASE-3/index.html). The Early Middle English poem *The Ormulum* is excluded. Texts translated from French and Latin originals are included.

Table 2.1. *A comparison of the incidence of* ne...not *predicted by the model in (5) with the frequency of* ne...not *attested in data from the PPCME2 corpus.*

| Period | Overall *ne* | Overall *not* | Total | Estimated *ne...not* | Observed *ne...not* |
|---|---|---|---|---|---|
| 1150–1250 | 713 (99.0%) | 284 (39.4%) | 720 | 281 (39.1%) | 277 (38.5%) |
| 1250–1350 | 633 (90.8%) | 539 (77.3%) | 697 | 490 (70.1%) | 484 (69.4%) |
| 1350–1420 | 279 (12.5%) | 2195 (98/1%) | 2238 | 274 (12.1%) | 236 (10.5%) |
| 1420–1500 | 32 (1.7%) | 1860 (99.3%) | 1874 | 32 (1.7%) | 18 (1.0%) |

b. But **noman** knoweth where he was buryed.
But no man knows where he was buried.
'But no-one knows where he was buried'
(CMMANDEV,41.1020)

c. and þow seist "whoeuere kepuþ my word schal **neuere** dye."
and you say "whoever keeps my word shall never die"
'and you say "whoever obeys my commands shall never die"'
(CMWYCSER,420.3488)

Clauses involving *ne* in combination with *bute(n)*, such as those given in (14), where the combination of *ne* and *buten* is interpreted with the meaning 'only' are also excluded. Jack (1978b) observes that the frequency of *ne* is much higher, and *ne...not* is much lower in these clauses than elsewhere.

(14) þer nis buten an godd þur hwam witerliche ha alle
there NEG-is but one God through whom undoubtedly they all
weren iwrahte ...
were made ...
'There is only one God through whom undoubtedly they were all made'
(CMKATHE,22.56)

Furthermore, contracted verbal proclitic negative forms, as illustrated in (15), such as *nys* 'not-is', *nylle* 'not-will' and *nyste* 'not-know' are excluded. These also pattern differently. Preterite present forms such as *nolde* 'would not' illustrated in (15a), and *nyste* 'knew not' illustrated in (15b) tend not to co-occur with *not* even in the fourteenth century. Instead they tend to retain *ne*.[3]

[3] Frequencies of contracted proclitic *n+V...not* from the PPCME2 with preterite present verbs are as follows: 1150–1250 *n+V* 69.2% (n=108/156); 1250–1350 *n+V* 56.5% (n=13/23); 1350–1420 *n+V* 72.7%(n=32/44). The totals include clauses with *n+V* and *n+V...not*. The frequency of *ne...not* does not increase over time with these predicates. One hypothesis to explain this divergent behaviour that those preterite presents that have a negative form beginning <*n-*> *nolde, nylle, noot, nyste* and so on, are lexicalised negative forms and hence exceptions to the Jespersen Cycle. Preterite present verbs are distinct from lexical verbs in many other respects too (see Warner 1993).

(15) a. for thei nolden make bateil in sabatis.
for they NEG-ought make battle on sabbath
'for they ought not fight on the sabbath'
(CMPURVEY,I,42.1833)

b. wherfore he nyst what forto done
for-that-reason he NEG-knew what for to do
'for that reason he did not know what to do'
(CMBRUT3,18.533)

The 'overall *ne*' and 'overall *not*' columns in Table 2.1 report the distribution of *ne* and *not* in the corpus data as a whole, whether *ne* and *not* appear in clauses alone, or in combination. This means that instances of *ne...not* are in effect counted twice within this model, once in the 'overall *ne*' column as an instance of *ne* and once in the 'overall *not*' column as an instance of *not*.

The redundant licensing model in (12) appears to provide a good fit to the PPCME2 data in Table 2.1, particularly in the two early periods 1150–1250 and 1250–1350. However, the fit of the data to the model progressively worsens over the course of the Middle English period, so that by the final subperiod 1420–1500, Frisch's model over-estimates the frequency of *ne...not* by a factor of 1.7 times its observed frequency. A chi-square test comparing the estimated and observed frequencies of *ne...not* in this period (shaded in Table 2.1) reports a significant difference between the estimated and observed incidence of *ne...not* ($\chi^2$(1df) = 3.868, p=.05).

In this form, the model takes no account of the distinction between adverbial *not* and negative marker *not*. Frisch's model assumes that all instances of *ne* and *not* are independent of each other. However, when used as an adverbial minimiser, *not* is a concordant negative item and must co-occur with the negative marker *ne*. The negative head *ne* will always be present when adverb *not* is present. The model in (12) will assume that more instances of *not* can appear independently of *ne* than is in fact the case if some instances of *not* are concordant adverbs rather than negative operators. Therefore, the model will underestimate the overall frequency of *ne...not* in Early Middle English. In order to implement the model more precisely, adverb *not* and negative operator *not* need to be separated, as in (16).

(16) Probability of *ne...not* overall =
(Probability *ne* is used (in $Neg^0$) x Probability *not* is used (in spec,NegP)) + Probability of adverb *not*
(adapted from Frisch 1997: 51)

(16) assumes that negative adverb *not* can be distinguished from negative operator *not* unambiguously. How this can be achieved is far from straightforward. Frisch (1997, 34ff) claims that pre-verbal *not* in examples like (17) is always an adverb.

(17) he hit naht ne wite
he it not NEG knew
'he did not know it' (CMTRINIT,79.1073)

He then asserts that the frequency at which adverb *not* preposes to a pre-verbal position is the same as for the adverb *never* (16 per cent). If 16 per cent of adverb *not* appear pre-verbally, then the overall frequency of adverb *not* is 'n (pre-verbal *not*) / 0.16'.

This calculation rests on a number of highly problematic assumptions. First, there is no justification for extrapolating the frequency of adverb *not* from the frequency of pre-verbal *never*. It is not at all clear that *not* and *never* are distributed in the same way in respect of the finite verb.[4] More importantly, Frisch's assumption that all pre-verbal instances of *not* are adverbs rather than negative markers relies on an assumption that verb movement is to a uniformly head-initial IP. However, Pintzuk (1999) observes a number of Old English clauses which require a head-final IP to derive them. Kroch and Taylor (2000) observe that similar head-final clause types persist into the Middle English period. These are precisely the clause types with a clause-final finite verb, such as (17) where we find pre-verbal *not*. Verb movement to the head of a head-final IP will result in pre-verbal *not* irrespective of whether *not* is an adverb or a negative marker. The frequency of pre-verbal *not* in the period 1150–1250 is entirely consistent with the frequency of INFL-final clauses estimated by Kroch and Taylor (2000) at around 5 per cent. In later periods there is no pre-verbal *not*. In any case, in these later periods (1350–1420 and 1420–1500), Frisch's redundant licensing model *over*estimates the frequency of *ne...not*. Therefore, excluding adverb *not* would not improve the fit of his model in those periods.

### 2.2.2 *Contextual Variation in the Distribution of Negative Forms*

Another shortcoming of Frisch's (1997) analysis is that it assumes a homogenous dataset, with the distribution of negative forms uniform across all texts and all contexts. The redundant licensing model in (12) should not only estimate the distribution of *ne...not* within the PPCME2 as a whole, but also model any variation in the distribution of *ne...not* across different contexts. Any variation in the frequency of *ne...not* should emerge via the intersection of different overall frequencies of *ne* versus Ø in $Neg^0$ and of *not* versus Ø in spec,NegP in each context.

Jack (1978a) identifies two groups of texts in the fourteenth-century data, based on differences in the frequencies of *ne*, *ne...not* and *not* used in these

[4] In Chapter 4, I argue that TP-adjoined *not* is highly circumscribed to particular contexts of contrastive polarity focus.

texts. He distinguishes a particular group of London texts which exhibit higher frequencies of *ne* and *ne...not* than do other texts. Jack highlights Chaucer's Boethius and the Canterbury Tales as representative of this dialect.

He observes that fourteenth- and fifteenth-century *ne* is more common in certain types of subordinate clauses than in main clauses. He highlights conditional clauses, such as *if*-clauses like (18), as a context for *ne*.

(18) if he ne hadde pitee of mannes soule, a sory song we myghten alle synge
if he NEG had pity on man's soul, a sorry song we might all sing
'if he didn't take pity on man's soul, we might all sing a sorry song'
(Parson's Tale 315, Jack (1978a, 61))

Jack (1978a) also distinguishes negative subordinate clauses which depend on a superordinate negative or interrogative clause, such as those given in (19), as particular contexts for *ne*, in contrast to other subordinate clauses.

(19) a. For ther nys no creature so good that hym ne wanteth somwhat of the perfeccioun of God
For there NEG-is no creature so good that he NEG needs something of the perfection of God
'For there is no creature so good that he doesn't need something of God's perfection'
(Melibee 1080, Jack (1978a, 60))

b. For what man is so sad or of so parfait welefulnesse, that he ne stryveth or pleyneth ayen the qualite of his estat?'
For what man is so satisfied or of so perfect happiness, that he NEG quarrels or complains about the quality of his circumstances?
'For who is so satisfied or perfectly happy that he doesn't quarrel or complain about his circumstances?'
(Boece II,pr.4,72–5, Jack (1978a, 60))

Can the differences between main and subordinate clauses observed by Jack (1978a) be modelled using Frisch's redundant licensing model? Tables 2.2 and 2.3 compare the frequencies of *ne...not* in main and subordinate clauses estimated under Frisch's model with the observed distribution of *ne...not* in each clause type in PPCME2 data. Cells are shaded where there is a statistically significant ($p \leq .05$) difference between estimated and observed frequencies of *ne...not*.

The estimated and observed frequencies correspond better in main clauses (Table 2.2) than in subordinate clauses (Table 2.3). While chi-square tests do not report any significant differences between estimated and observed frequencies of *ne...not* in main clauses, the differences for 1420–1500 are highly significant for subordinate clauses when they are considered separately ($\chi^2(1df)=6.67$, $p \leq .01$).

Table 2.2. *Comparison of the frequency of* ne...not *estimated by the redundant licensing model and the incidence of* ne...not *observed in PPCME2 data. Main clauses only.*

| Period | Overall *ne* | Overall *not* | Total | Estimated *ne...not* | Observed *ne...not* |
|---|---|---|---|---|---|
| 1150–1250 | 243 (98.8%) | 155 (63.0%) | 246 | 155 (62.9%) | 152 (61.8%) |
| 1250–1350 | 365 (89.0%) | 365 (89.0%) | 410 | 325 (79.3%) | 329 (77.4%) |
| 1350–1420 | 111 (10.6%) | 1043 (99.7%) | 1046 | 111 (10.6%) | 108 (10.3%) |
| 1420–1500 | 13 (1.4%) | 939 (99.9%) | 940 | 13 (1.4%) | 12 (1.3%) |

Table 2.3. *Comparison of the frequency of* ne...not *estimated by the redundant licensing model and the incidence of* ne...not *observed in PPCME2 data. Subordinate clauses only.*

| Period | Overall *ne* | Overall *not* | Total | Estimated *ne...not* | Observed *ne...not* |
|---|---|---|---|---|---|
| 1150–1250 | 470 (99.2%) | 129 (27.2%) | 474 | 128 (27.0%) | 125 ( 26.4%) |
| 1250–1350 | 268 (93.3%) | 174 (60.6%) | 287 | 162 (56.6%) | 155 (54.0%) |
| 1350–1420 | 168 (14.1%) | 1152 (96.6%) | 1192 | 162 (13.6%) | 128 (10.7%) |
| 1420–1500 | 19 (2.0%) | 921 (98.0%) | 934 | **19 (2.0%)** | **6 (0.6%)** |

The fit of Frisch's model is worse for the two types of subordinate clauses that Jack (1978a) claims favour *ne* than it is for other clauses. Table 2.4 compares estimated and observed distributions of *ne...not* in *if*-clauses, and Table 2.5 does the same for subordinate clauses which are dependent on a negative clause. Again cells are shaded where there is a statistically significant ($p \leq .05$) difference between estimated and observed frequencies of *ne...not*. For the period 1420–1500, figures are too small to perform $\chi^2$-tests.

Many of the figures in these tables are small, often providing insufficient data to establish if the differences reported between the estimated and observed frequencies of *ne...not* are statistically significant. However, even allowing for the greater potential for errors or anomalies in these small datasets than in the larger main and subordinate clause datasets presented in Tables 2.2 and 2.3, it seems that in subordinate environments that Jack identifies as contexts favouring *ne* in fourteenth- and fifteenth-century data, the fit of Frisch's model is poorer than in main clause data from the same time period. Frisch's model proves a consistently poor fit to fourteenth- and fifteenth-century data from a range of subordinate clause types. Furthermore, the divergence between the estimated frequency of *ne...not* and the frequency of *ne...not* observed in the PPCME2 data always follows the same direction – that is, the estimate of *ne...not*'s frequency is consistently higher than the observed frequency in subordinate clauses, and the degree of

Table 2.4. *Comparison of the frequency of* ne...not *estimated by the redundant licensing model and the incidence of* ne...not *observed in PPCME2 data.* If-*clauses only.*

| Period | Overall *ne* | Overall *not* | Total | Estimated *ne...not* | Observed *ne...not* |
|---|---|---|---|---|---|
| 1150–1250 | 43 (100.0%) | 10 (23.6%) | 43 | 10 (23.6%) | 10 ( 23.6%) |
| 1250–1350 | 20 (87.0%) | 11 (47.8%) | 23 | 10 (41.6%) | 8 (34.7%) |
| 1350–1420 | 14 (17.9%) | 69 (88.5%) | 78 | 12 (15.8%) | 5 (6.4%)[5] |
| 1420–1500 | 3 (5.6%) | 52 (96.3%) | 54 | 3 (5.4%) | 1 (1.9%) |

Table 2.5. *Comparison of the frequency of* ne...not *estimated by the redundant licensing model and the incidence of* ne...not *observed in PPCME2 data. Clauses dependent on a negative clause only.*

| Period | Overall *ne* | Overall *not* | Total | Estimated *ne...not* | Observed *ne...not* |
|---|---|---|---|---|---|
| 1150–1250 | 36 (100.0%) | 3 (8.3%) | 36 | 3 (8.3%) | 3 ( 8.3%) |
| 1250–1350 | 25 (92.3%) | 8 (29.6%) | 27 | 7 (27.3%) | 6 (22.2%) |
| 1350–1420 | 22 (28.6%) | 63 (81.8%) | 77 | 18 (23.4%) | 8 (10.4%)[6] |
| 1420–1500 | 5 (10.4%) | 44 (91.7%) | 48 | 5 (9.5%) | 1 (2.1%) |

difference between estimated and observed frequencies of *ne...not* consistently increases over the course of Middle English as *ne* is lost, being much greater in the later sub-periods when the overall frequency of *ne* is lower and the overall frequency of *not* is higher. These systematic patterns of poor fit between the model and the data indicate that the reason for the poor fit of Frisch's model to these data is to do with the way his model is constructed, rather than to do with idiosyncrasies or anomalies in the PPCME2 data.

### 2.2.3 *Conceptual Issues in Modelling the Loss of* ne *and Introduction of* not *as Diachronically Independent Processes*

In addition to the empirical problems noted in Sections 2.2.1 and 2.2.2, deriving *ne...not* via the intersection of two independent changes is conceptually problematic in two major respects. It assumes that the loss of

[5] The difference between estimated and observed frequencies of *ne...not* is reported as statistically non-significant here by a chi-square test. We observe 5 clauses with *ne...not* and 73 clauses without *ne...not*, whereas the model estimates 12 clauses with *ne...not* and 66 clauses without *ne...not*. Performing a $\chi^2$ test on this 2x2 matrix gives a result of $\chi^2$ (1df) = 3.235, p=.07.

[6] The difference between estimated and observed frequencies of *ne...not* is statistically significant here. We observe 8 clauses with *ne...not* and 69 clauses without *ne...not*, whereas the model estimates 18 clauses with *ne...not* and 59 clauses without *ne...not*. Performing a chi-square test on this 2x2 matrix gives a result of chi-square (1 df) = 4.627, p=.03.

*ne* is independent of the introduction of *not*. The co-occurrence of the two forms in the *ne...not* construction is merely a historical accident – two independent processes happen to overlap. This might be plausible if bipartite negative forms were found only in a small number of languages. However, the Jespersen Cycle occurs and recurs in many languages (see Dahl (1979), Willis et al. (2013), Bernini and Ramat (1996)).[7] It seems most plausible that there are syntactic or structural factors underlying the Jespersen Cycle that cause this sequence of changes to reccur cross-linguistically, and which the redundant licensing model does not capture.

Furthermore, while Frisch's quantitative model maintains the independence of the two changes, his syntactic model implies that the changes affecting spec,NegP and $Neg^0$ are in fact related, rather than truly independent. Frisch acknowledges that a syntactic analysis of the Jespersen Cycle in terms of two independent changes affecting *ne* and *not* is insufficiently constrained. Conceptually, two independent changes give rise to four (2x2) options for the expression of negation, as illustrated in (20).

(20)

| $Neg^0$ | spec,NegP |
|---|---|
| *ne* | Ø |
| Ø | *not* |

(20) gives rise to four structural options, three of which are overt and attested in the Middle English data – *ne, ne...not* and *not* – and one in which a null operator in spec,NegP and a null head in $Neg^0$ co-occur, so that negation is not marked by any overt morpheme. However, negation must be morphologically identified in order to distinguish negative and affirmative clauses. Frisch proposes an output filter on the derivation of negative clauses in order to rule out negative clauses with non-overt head and non-overt specifier – clauses in which negation would not be morphologically marked. Incorporating this filter into the syntactic account is an implicit admission that the changes affecting *ne* and *not* are not entirely independent, but are related, so that a non-overt head requires an overt specifier and a non-overt specifier requires an overt head. Frisch's syntactic and quantitative models are inconsistent. While his syntactic argument invokes such an output filter, as a post-hoc filter on derivations, this is difficult to implement in his quantitative model; thus his quantitative model seems not to take it into account.[8] In ruling out derivations in which both $Neg^0$ and spec,NegP are non-overt, Frisch's syntactic model actually structures the change as in (21).

7 Although there are some differences, for detailed discussion of which see van der Auwera (2009) and Willis (2013).

8 The analyses in Roberts and Roussou (2003) and Zeijlstra (2004) also face these issues and posit various conditions on the morphological identifiability of negation.

(21)

<table>
<tr><th>Neg$^0$</th><th>spec,NegP</th><th></th></tr>
<tr><td rowspan="2">ne</td><td>Ø</td><td>= ne</td></tr>
<tr><td rowspan="2">not</td><td>= ne...not</td></tr>
<tr><td>Ø</td><td>= not</td></tr>
</table>

This model structures the Jespersen Cycle in a different way. What this model says is that, in order to understand the changes within the Jespersen Cycle, we have to treat the change from *ne* to *ne...not* as one change, and the change from *ne...not* to *not* as a separate change. However, this is difficult to formalise in a NegP licensing approach without invoking output filters on syntactic derivations, which seem conceptually unattractive for reasons of economy. Frisch's model allows the derivation of clause structures which are never attested and thus have to be ruled out post hoc.

### 2.2.4 *The Distributional Independence of* ne, ne...not *and* not

Table 2.6 presents the overall distributions of *ne, ne...not* and *not* in the PPCME2 within each of the four clausal contexts discussed in Section 2.2.2. These data support the view that *ne*, *ne...not* and *not* are independent. Figure 2.1 illustrates the first two contexts in Table 2.6 – main and subordinate clauses. In all periods, *not* is distributed in the same way across both clause types, but *ne* is more frequent in subordinate clauses and *ne...not* more frequent in main clauses. Looking particularly at the periods 1350–1420 and 1420–1500, we see that as *ne* is lost so are the differences between main and subordinate clauses. The frequencies of *ne...not* also become increasingly similar across all grammatical environments. These data show that clause-type constraints on *ne...not* weaken over time as *ne* is lost while not remains unaffected by this change. These different patterns are difficult to interpret without treating *ne, ne...not* and *not* as three independent forms. The graph suggests that there are clause-type constraints on competition between *ne...not* and *ne*, but not on competition between *ne...not* and *not*. I argue in Section 2.3.4 that this is essentially correct and that – once we structure the change appropriately – the clause-type constraints on all three forms remain constant within a logistic regression model.

There are two ways in which *ne...not* may be distinct from *ne* and from *not*. The distribution of *ne* in bipartite *ne...not* may differentiate it from *ne* which negates a clause on its own. If so, there is a change in the distribution of *ne* in the transition from *ne* to *ne...not*. Alternatively, the distribution of *not* in bipartite *ne...not* may differentiate it from *not* which negates a clause on its own. If so, there is a change in the distribution of *not* in the transition from *ne...not* to *not*. It is equally possible that both of these possibilities are realised in the data. So is the distinct distribution of *ne...not* due to a distinction between two types of *ne* or between two types of *not* (or both)?

Table 2.6. *The distribution of ne, ne...not and* not *in four clause types. Data from the PPCME2.*

| Period | Main clauses: ne | ne...not | not | Total | Subordinate clauses: ne | ne...not | not | Total | *if*-clauses: ne | ne...not | not | Total | Dependent on a neg-clause: ne | ne...not | not | Total |
|---|---|---|---|---|---|---|---|---|---|---|---|---|---|---|---|---|
| 1150–1250 | 91 (37.0%) | 152 (61.8%) | 3 (1.2%) | 246 | 279 (68.3%) | 125 (26.4%) | 4 (1.0%) | 408 | 33 (76.7%) | 10 (23.6%) | 0 – | 43 | 33 (91.7%) | 3 (8.3%) | 0 – | 36 |
| 1250–1350 | 48 (11.3%) | 329 (77.4%) | 48 (11.3%) | 425 | 87 (33.1%) | 161 (61.2%) | 15 (5.7%) | 263 | 12 (52.2%) | 8 (34.7%) | 3 13.0%) | 23 | 19 (70.4%) | 6 (22.2%) | 2 (7.4%) | 27 |
| 1350–1420 | 3 (0.3%) | 108 (10.3%) | 935 (89.4%) | 1046 | 17 (1.6%) | 128 (12.2%) | 905 (86.2%) | 1050 | 9 (11.5%) | 5 (6.4%) | 64 (82.1%) | 78 | 14 (18.2%) | 8 (10.4%) | 55 71.4%) | 77 |
| 1420–1500 | 1 (0.1%) | 12 (1.3%) | 927 (98.6%) | 940 | 7 (0.8%) | 6 (0.6%) | 821 (98.4%) | 834 | 2 (3.7%) | 1 (1.9%) | 51 (94.4%) | 54 | 4 (8.3%) | 1 (2.1%) | 43 (89.6%) | 48 |

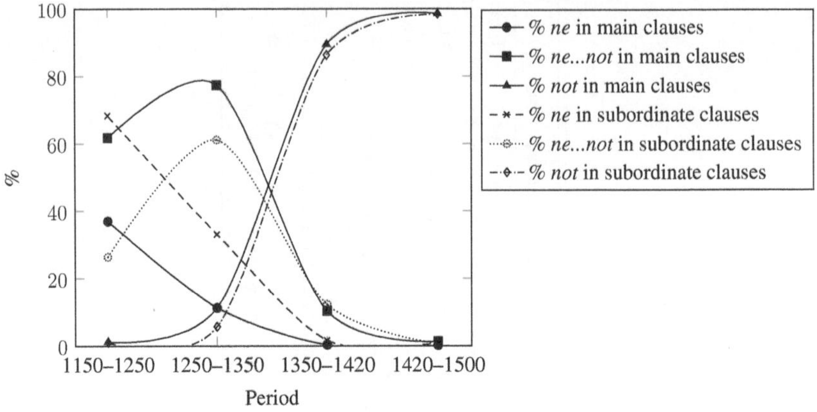


Figure 2.1 The distributions of *ne, ne...not* and *not* in main and subordinate clauses

The distinctiveness of *ne...not* does not itself necessarily imply a distinction between two types of *ne* at successive stages of the Jespersen Cycle. So we cannot distinguish between the two types of syntactic analysis outlined earlier in (9) and (10) without further analysis of the source of *ne...not*'s distinctiveness. The distinctiveness of *ne...not* may arise because there are two types of *not* – for example, *not* may be an adverbial when it co-occurs with *ne*, but a grammaticalised negative marker when it negates a clause independently. If some or all of the instances of *not* in the *ne...not* construction are adverbs rather than negative markers, then the distributional independence of *ne...not* and *not* in the corpus data might be accounted for by different constraints upon the distributions of the two different types of *not*, without also having to postulate two types of *ne*. Such an account is consistent with the model of Jespersen Cycle in (22).

| | | | | | |
|---|---|---|---|---|---|
| (22) | Stage one: | Op[NEG] | + *ne* | | |
| | Stage two: | Op[NEG] | + *ne* | + adv *not* | (*not* is an adverb) |
| | Stage three: | *not*[NEG] | + *ne* | | (*not* is a neg marker) |
| | Stage four: | *not*[NEG] | | | (*not* is a neg marker) |

Alternatively, the grammaticalisation of *not* might coincide with the loss of *ne*. A way to capture this relationship is (23), in which the locus of negation (here represented as [NEG]) changes from *ne* to *not* at stage three of the cycle.

| | | | | |
|---|---|---|---|---|
| (23) | Stage one: | *ne* [NEG] | | |
| | Stage two: | *ne* [NEG] | + adv *not* | (*not* is an adverb) |
| | Stage three: | *not*[NEG] | | (*not* is a neg marker) |

Here, the grammaticalisation of *not* is a consequence of the loss of *ne*. The loss of *ne*[NEG] creates a functional need for another negative marker, prompting grammatical reanalysis of the negative adverb as a negative marker. Both these models require that *ne* is a single lexical item throughout the Jespersen Cycle, predicting continuity in the distribution of *ne* both when it is used independently and when it co-occurs with *not*. In Section 2.3, I argue that changes in the distribution of *ne* and *not* provide empirical evidence to distinguish between these possible models.

### 2.2.5 *Summary*

This section showed that an analysis which conceptualises the loss of *ne* and the introduction of *not* as two independent changes faces both conceptual and empirical problems. These are summarised as follows:

1. This approach makes null realisations of $Neg^0$ and spec,NegP available, but requires an output filter on syntactic derivations to rule out the appearance of clauses in which negation is not overtly marked (clauses in which both spec,NegP and $Neg^0$ are null).
2. This model does not predict the different frequencies of *ne* and *ne...not* observed in subordinate clauses in the PPCME2. These differences in the frequencies of *ne* and *ne...not* do not fall out of differences in the overall frequencies of either *not* or *ne* in a model which treats *ne* as a single lexical item whose syntax is the same throughout the Jespersen Cycle.
3. The overall frequencies of *ne* and *not* are not independent of each other. Compare Tables 2.2–2.5 – in contexts where the overall frequency of *ne* is higher and the overall frequency of *not* is lower and vice versa. In fact, Figure 2.1 indicates that *ne*...*not* and *not* are independent of each other while *ne* and *ne*...*not* compete.

In order to ascertain whether the independence of *ne...not* comes from two distributionally independent forms of *ne* as the model in (10) predicts, we need to compare the constraints on *ne* when it occurs independently and *ne* which co-occurs with *not*. Different constraints on the distribution of *ne* at each of these stages would imply that there are two distinct forms of *ne* in the data in competition with each other, whereas continuity in the constraints on *ne* at each stage could suggest a single form of *ne* throughout the cycle. The same applies to *not*.

In Section 2.3, I test these predictions using logistic regression analyses to examine the constraints on *ne, ne...not* and *not* during the Jespersen Cycle. We will see that Frisch's (1997) model cannot structure the clause-type constraints on *ne...not* in a way that is consistent with the model of grammatical competition set out in Kroch (1989), providing another reason to reject Frisch's model.

## 2.3 An Alternative Model of the Jespersen Cycle: Evidence from Change in Progress

### 2.3.1 *Grammar Competition and the Constant Rate Effect*

In Section 2.2, we saw distributional differences between *ne*, *ne...not* and *not* in the fourteenth and fifteenth centuries. Do the clause-type constraints on *ne*, and the distributional differences between *ne* and *ne...not* change during Late Middle English (LME), or is competition between *ne*, *ne...not* and *not* structured in the same way throughout Middle English? Kroch's (1989) grammar competition model provides a theoretical and methodological framework through which we can approach this question. The grammar competition model makes clear predictions about how parametric change should pattern across time and across grammatical contexts. Thus we can evaluate different models of the Jespersen Cycle, such as Frisch's redundant licensing model, according to how well they structure change in progress as grammatical competition. Models of change in progress therefore provide evidence to choose between syntactic analyses of the Jespersen Cycle.

Kroch's (1989) hypothesis is that grammatical change proceeds at the same rate in each context where it occurs. This is the Constant Rate Hypothesis (Kroch, 1989, 205ff) or Constant Rate Effect (Kroch, 1994, 181). The converse of claiming that the rate of change is the same in each context is to say that contextual constraints on the occurrence of competing forms are constant throughout the change. In this model, the effect of context on the variation between a pair of forms is established at the actuation of the change. As change progresses, only the overall frequencies of the forms change, not the contextual constraints on their use. Kroch explains:

> A constant rate of change across contexts is mathematically equivalent to a fixity of contextual effects, in direction and size, across time periods. Thus if a study reports a series of multivariate [logistic regression] analyses for different time periods, and the contextual effects are constant across these analyses, the rate of change of each context measured separately would necessarily be the same.
> (Kroch, 1989, 205)

So can we structure the Jespersen Cycle as a series of independent changes, all of which proceed at a constant rate in all contexts?

There is an important distinction to be made here between the frequency of a form in a particular context and the constraining effect that the context has on the likelihood or probability of a form occurring there. We cannot estimate the effect of contextual constraints directly by calculating differences in the frequency of a form across two or more contexts. This is because the progress of diachronic change over time is not linear. Instead it follows a logistic curve, produced by the function in (24), where $p$ is the frequency of

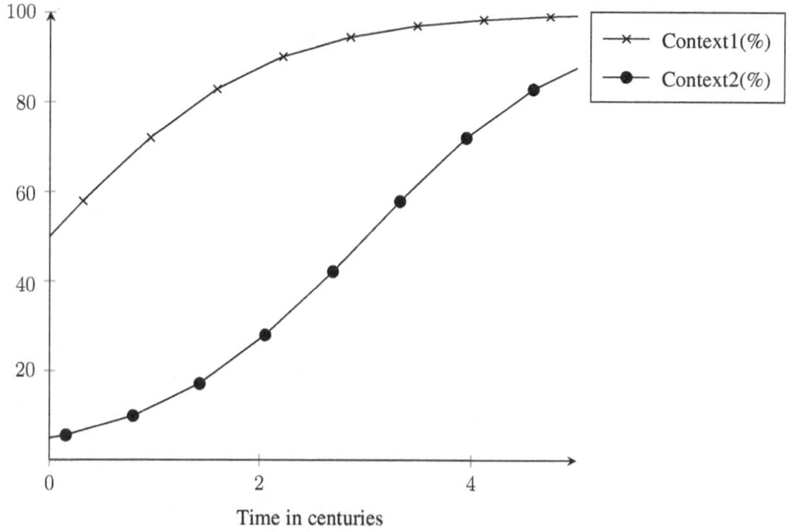


Figure 2.2 Parallel logistic curves with the same slope, but different starting frequencies (intercepts)

the advancing form; $e$ is a constant, Euler's number, 2.718; $t$ represents time; $k$ represents the intercept parameter, the frequency of the advancing form at the beginning of the change, time t=0; and $s$ is the slope of the curve.

(24) The logistic function (from Kroch (1989, 202, ex.1): $p = \left(\frac{e^{k-st}}{1+e^{k-st}}\right)$

Consequently, the frequencies of a particular competing form in two or more contexts will appear to diverge and converge over time, even if the contextual constraints on the form (different values of intercept $k$ in different contexts) remain constant throughout the change, and the slope of the change ($s$) is the same in all contexts. Figure 2.2 illustrates this.

In order to compare the strength of contextual constraints at different points within a change, we need logistic regression analyses to 'flatten out' the logistic curves, so we can compare the contextual constraints on a form at different points on the curve, and see whether the probabilistic effect each context has upon the occurence of a variant form – the likelihood that a variant will occur in that context – remains the same throughout the change. Logistic regression allows us to compare the probabilistic constraints on a form at any two or more points during a change irrespective of different overall frequencies of forms in the datasets. It also allows us to estimate the probabilistic effect of several constraints upon a form simultaneously, and establish whether those constraints are independent of each other or whether they interact.

### 2.3.2 *Mixed-Effects Regression Models*

The most widely practised logistic regression analysis technique in variationist sociolinguistics uses a form of VARBRUL (see Sankoff and Labov (1979), Tagliamonte (2006)) to estimate probabilistic factor weights. Where there is variation between a pair of forms, these factor weights represent the likelihood that one form rather than the other will appear in each context. However, VARBRUL analyses are problematic in many ways, particularly in respect of corpus data, as (for example) Johnson (2008), Tagliamonte and Baayen (2012), Gries (2015b) observe. VARBRUL analyses operate under the assumption that each token of a particular variable is independent of all the other tokens in the dataset – that each informant provides a single token to the analysis. Johnson (2008) observes that the number of Type I errors (false reports of statistical significance, overestimation of p-values) within a VARBRUL analysis increases the more this assumption is violated. He demonstrates using a constructed dataset that the significance of a speaker variable (gender) increases the more tokens each individual speaker contributes to the dataset. The further we go from the idealised one observation per speaker, the higher the rate of Type I errors becomes.

Clearly, in corpus data the assumption that each token is independent of every other token is false, so corpus data violate the assumptions on which VARBRUL and other generalised linear regression models are based. In written corpus data, it is obvious that individual writers typically contribute more than one token or observation per variable. Particularly in the case of the diachronic corpora under investigation here, a fairly limited number of corpus texts provide multiple tokens within the dataset. While efforts were made to balance diachronic corpora such as the PPCME2 (Kroch and Taylor, 2000) across texts and registers, the composition of the corpus is largely dictated by the historical data that survive to us. Historical corpus data are not products of structured sampling, such as is usually carried out in sociolinguistic studies to ensure different groups contribute roughly equal numbers of tokens to the dataset. In the PPCME2, the numbers of negation tokens contributed by each corpus text are highly uneven. Larger texts provide more data for analysis, potentially skewing the model towards them in a way that might result in Type I errors. Within a VARBRUL analysis it is difficult to discern whether a statistically significant effect is significant because it is a particularly strong effect in an individual text or whether the effect generalises across all texts in the corpus equally.

A more nuanced model needs to take into account the intermediate level of organisation within the corpus – that tokens are nested under texts, and that the texts comprise the corpus (see also Gries (2015a)). Patterns of variation within texts and between texts interact in the model. The Constant Rate Effect, when applied at the level of the individual texts implies that constraints on variation within an individual text should also hold across all other texts in the corpus as a whole. The statistical analysis therefore needs to take into account the ways in which these two levels of variation – within and across texts – interact within the dataset. It is possible to include 'text' as a factor within a VARBRUL analysis to get some estimate of the differences between individual corpus texts. However, VARBRUL does not handle interactions between factors, meaning that questions of whether grammatical factors pattern uniformly across texts, or differ within individual texts are difficult to investigate. VARBRUL assumes that all factors within the analysis are independent of each other.

Differences between texts and differences between grammatical environments are of different sorts. Grammatical environments are predictors. The results of our analysis of one dataset can make predictions about the behaviour of grammatical constraints more generally. For example, under the Constant Rate Effect, we can estimate the likelihood of finding *not* in a particular context in twelfth-century data and, from this, make a prediction about the likelihood of finding it in the same context in the fourteenth century. We cannot do this with individual texts. The likelihood of finding a form in a particular text does not predict the likelihood of that form in another text, simply because the factors that underlie differences between texts are complex interactions of geographical, stylistic and author variables, not all of which can be modelled within the analysis. Differences between texts may simply be idiosyncratic. Hence, we require a mixed-effects generalised linear regression analysis (Bates, 2010) which incorporates grammatical constraints as fixed effects (or predictors) and corpus text as a random effect. A mixed-effects model can model the interaction between these different effects, taking into account the shape and structure of the dataset within the model. Taking into account variation across individual texts provides more reasonable probabilistic estimates of grammatical constraints, both in terms of effect sizes and p-values, than in only fixed effects models. Typically effect sizes and p-values for predictors shrink within mixed-effects models. Mixed-effects models may overstate the random effects and underestimate the fixed effects, resulting in Type II errors (false reports of statistical non-significance, or underestimation of effect sizes or p-values).

Within a mixed-effect regression model, there are two dimensions in which texts might differ from each other, either or both of which might cause variation to be unevenly distributed within the corpus data. First, the overall frequency of a particular variant might differ across individual texts (differences in the intercept parameter). Second, the strength of the grammatical constraints upon a particular variant may differ across texts (the slope parameter). Mixed-effect regression models provide an opportunity to investigate intercept or slope differences between texts in more detail, in ways which might allow us to group texts together, for example according to independent variables of style, register or dialect which could then be incorporated into the model as predictors.

When we incorporate text as an intermediate level in our dataset, we can test the application of the Constant Rate Effect at the level of individual corpus texts (providing those corpus texts yield sufficient data for analysis). At this level, the Constant Rate Effect predicts that grammatical constraints on a form will pattern the same way across all the corpus texts – that is, that differences between the texts will be differences in the overall frequencies of competing forms (the intercept) rather than differences in the constraints upon those competing forms (the slope). In order to test this, we require comparison between two types of mixed-effects models, those which allow intercepts to vary across texts, and those which allow both intercepts and slopes to vary across texts. We can then perform a comparison of these two models using an analysis of deviance as described by both Tagliamonte and Baayen (2012) and Gries (2015b), to determine if the fit of the two models is significantly different. If the slopes for each grammatical constraint do not differ significantly across texts, we may conclude that our model requires only variable intercepts for each text, i.e. that it is one in which the Constant Rate Effect holds and grammatical constraints pattern the same way in each text.

Rbrul (Johnson, 2008) is one way to produce mixed-effects models, but it does not yet allow random slopes – only random intercepts – for random effects. Therefore, the following analysis is carried out in R using the LMER package (Bates, 2010).[9] This allows random slopes and random intercepts separately or in combination, reporting random effects as deviations from the mean intercept or slope parameters. It also allows interactions between fixed effects (or predictors) to be modelled. The analyses in this section include a fixed effect, or predictor, 'clause type' with levels main clauses, subordinate clauses, negative second conjuncts, if-clauses (split into conditional if-clauses and counterfactual if-clauses) and negative clauses which are themselves dependent on a superordinate negative clause. The results of the model are reported not as factor weights estimated via sum coding as is typical in VARBRUL analyses and interpreted with respect to the mean (with weights >.5 said to be favouring and weights <.5 said to be disfavouring of a particular variant), but on the log-odds scale using treatment coding and centred around zero (with positive log-odds values favouring a particular variant and negative log-odds values disfavouring it). Treatment coding takes one of the levels of the fixed effect, in this instance 'main clauses', as the reference level. The log-odds value reported by the analysis for each of the levels in the analysis is a probabilistic estimate of the difference between each predictor level and the reference level, estimated in logits. Consequently, the favouring or disfavouring effect is estimated for each predictor level in log-odds relative to the reference level. Tagliamonte and Baayern (2012: 12) observe that treatment coding produces better estimates when dealing with

9 For information about mixed-effects models in R see Baayen (2008), Baayen et al. (2008) and Gałecki and Burzykowski (2013).

highly unbalanced datasets (such as those based on diachronic corpora as here), and that it allows for more transparent estimates and interpretation of interactions between effects (both fixed and random).

The analysis also reports a p-value for each level of each predictor. This is an estimate of the statistical significance of any difference between the distribution within that particular context (or level) when compared to the distribution within the reference level (in this instance, the reference level is 'main clauses'). As is typical, p-values of $<.05$ are regarded as statistically significant, although Bonferoni corrections may be applied to analyses in which the effects of several contextual constraints are estimated simultaneously. The ability to construct and test complex models of variation, in which several different constraints are modelled simultaneously, both independently of each other and in interactions with each other, is another advantage of logistic regression analyses. This will become relevant later, as I construct a model of the Jespersen Cycle, and examine the interaction of grammatical and functional constraints on the distribution of negative forms in Chapter 6.

### 2.3.3 *Mixed-Effects Regression and the Redundant Licensing Model*

Using a series of separate logistic regression analyses performed on the data for each of the PPCME2 sub-periods (1150–1250, 1250–1350, 1350–1420, 1420–1500), we can test Frisch's (1997) model to see if the contextual effect of clause type on *ne* and *not* remains constant throughout the change. Frisch's model assumes two processes of competition within the Jespersen Cycle – *ne* versus Ø in $Neg^0$ and *not* versus Ø in spec,NegP – hence two sets of logistic regression analyses are required. The first set of analyses report the likelihood of $Neg^0$ being realised overtly as *ne* (*ne* is the application value) in each sub-period. The analysis includes clause type as a fixed effect or predictor, and corpus text as a random effect. The log-odds take 'main clauses' as the reference level. The results of these analyses are presented in Table 2.7. Where there are gaps in the table, these are contexts in which there are too few data to include that context in the regression analysis. Contexts in which there are fewer than three tokens of any variant are excluded from the regression analyses on the grounds that such small numbers of tokens are insufficient to produce reliable estimates of factor weights or constraints. Throughout the statistical analysis, only texts which exhibit variation between the variants under discussion are included. Invariant texts are excluded.

The greater the effect size, the more likely *ne* is to appear. Positive effect sizes appear in contexts that favour *ne* more than the reference level – *ne* is more likely to appear in these contexts than it is in main clauses. Conversely, negative effect sizes appear where *ne* is less likely than it is in main clauses. In all contexts except subordinate clauses, the effect size increases over time, so *ne* becomes more highly favoured as the change

Table 2.7. *Clause-type constraints on* ne.

| Period[10] | Subordinate clauses | | Neg Conj | | Conditional clauses | | Counterfactual *If*-clauses | | Dependent on neg-clause | |
|---|---|---|---|---|---|---|---|---|---|---|
| | Effect | p | Effect | p | Effect | p | Effect | p | Effect | p |
| 1250–1350 | -0.22 | .52 | 0.31 | .63 | -1.15 | .11 | – | – | -1.00 | .23 |
| 1350–1420 | 0.15 | .37 | 0.27 | .52 | -0.04 | .93 | 1.72 | .03 | 0.99 | .005 |
| 1420–1500 | -0.037 | .94 | 1.13 | .20 | 0.10 | .93 | 3.46 | .006 | 1.83 | .005 |

progresses. In LME, *ne* becomes significantly more likely to appear in counterfactuals and clauses dependent on a negative clause than in main clauses. Two observations point to this conclusion: first, the p-values in these two contexts; second, an increase in the effect size for each of these contexts over time. Although the redundant licensing model fits main clauses, the fit of the model to various subordinate clause types progressively worsens over time, as it fails to predict these emerging contextual constraints. Thus these regression results provide evidence against a constant rate of change across all clause types. The loss of *ne* in the redundant licensing model proceeds at a faster rate in main clauses than some subordinate clause types. Therefore, the redundant licensing model does not model the distribution of *ne* in a way that accords with Kroch's grammar competition model.

The second set of analyses report the likelihood of spec,NegP being realised overtly as *not* (*not* is the application value) in each sub-period. The results are presented in Table 2.8. Table 2.8 shows that the various subordinate clause types disfavour *not* – negative effect sizes indicate that *not* is less likely to occur in subordinate clauses than in main clauses. The typically small p-values indicate that the likelihood of finding *not* in each subordinate clause type differs significantly from the likelihood of finding it in main clauses. With the exception of the period 1250–1350, the effects are largely consistent across time, in both size and significance. Contexts providing more tokens to the analysis exhibit greater consistency in effect sizes (for example subordinate clauses). Unlike in the distribution of *ne* (Table 2.7), the effect sizes do not change over time. This consistency in effect sizes is evidence that the rate at which *not* is introduced is constant across all clause types from 1150 to 1500.[11]

[10] The period 1150–1250 is excluded from the table because *ne* is near categorical at this period.
[11] The effect sizes for the period 1250–1350 are typically somewhat lower than the other periods. The small number of texts in this period, and the high degree of difference between them probably results in the random effect of 'text' having a greater effect on the variation than the predictor 'clause type'.

Table 2.8. *Clause-type constraints on* not.

| Period | Subordinate clauses | | Neg Conj | | Conditional clauses | | Counterfactual *If*-clauses | | Dependent on neg-clause | |
|---|---|---|---|---|---|---|---|---|---|---|
| | Effect | p | Effect | p | Effect | p | Effect | p | Effect | p |
| 1150–1250 | -1.75 | .001 | -1.763 | .001 | -1.79 | .001 | – | – | -3.10 | .001 |
| 1250–1350 | -0.55 | .001 | -0.549 | .41 | -2.22 | .001 | – | – | -1.74 | .001 |
| 1350–1420 | -1.96 | .01 | – | – | -3.49 | .001 | -6.08 | .001 | -4.64 | .001 |
| 1420–1500 | -1.36 | .11 | – | – | – | – | -5.56 | .001 | -3.67 | .001 |

The results of the regression analyses reported in Tables 2.7 and 2.8 show that clause type has a statistically significant effect on both the competition for the $\text{Neg}^0$ position and the competition for the spec,NegP position in LME after 1350, but that the clause-type constraints on *ne* ($\text{Neg}^0$) and *not* (spec,NegP) pattern in opposite directions. *Ne* is favoured and *not* disfavoured in counterfactual clauses and clauses dependent on a negative. This suggests that there is in fact interaction – or competition – between *ne* and *not* on some level that deriving their frequencies in the data through two independent changes does not capture.

The constraints on *not* in spec,NegP (Table 2.8) are highly consistent over time, in the way that the Constant Rate Effect predicts. The consistency of these constraints over time indicates that the introduction of *not* proceeds at the same rate in all contexts, and by extension that these constraints are unaffected by the transition from stage two *ne...not* to stage three *not*. Therefore, we cannot appeal to changes in the distribution or syntax of *not* to explain the poor fit of the redundant licensing model. On the other hand, the distribution of *ne* in $\text{Neg}^0$ (Table 2.7) is less consistent. The likelihood of *ne* occurring in several clause types seems to vary systematically over time in a way which the Constant Rate Effect does not predict. The behaviour of *ne* rather than *not*, then, is responsible for the poor fit of the redundant licensing model in LME.

### 2.3.4 *An Alternative Grammatical Competition Model*

Tables 2.9–2.11 treat *ne, ne...not* and *not* as independent forms, involved in three processes of competition. Table 2.9 presents constraints on the competition between *ne* and *ne...not*, Table 2.10 the constraints on competition between *ne* and *not*, and Table 2.11 the constraints on competition between *ne...not* and *not*.

Tables 2.9 and 2.10 indicate that the constraints on competition between *ne* and *ne...not* and between *ne* and *not* are the same. Clause type is a highly significant constraint on both changes, and the effects of the various clause types are largely consistent from sub-period to sub-period throughout

Table 2.9. *Regression analysis identifying constraints on competition between* ne *and* ne...not.

| Period | Subordinate clauses | | Neg Conj | | Conditional clauses | | Counterfactual *If*-clauses | | Dependent on neg-clause | |
|---|---|---|---|---|---|---|---|---|---|---|
| | Effect | p | Effect | p | Effect | p | Effect | p | Effect | p |
| 1150–1250 | -1.75 | .001 | -1.76 | .001 | -1.79 | .001 | – | – | -3.10 | .001 |
| 1250–1350 | -0.57 | .02 | -0.71 | .29 | -2.04 | .001 | – | – | -1.78 | .001 |
| 1350–1420 | -2.19 | .02 | -1.78 | .23 | -4.09 | .001 | – | – | -5.11 | .001 |

Table 2.10. *Regression analysis identifying constraints on competition between* ne *and* not.

| Period | Subordinate clauses | | Neg Conj | | Conditional clauses | | Counterfactual *If*-clauses | | Dependent on neg-clause | |
|---|---|---|---|---|---|---|---|---|---|---|
| | Effect | p | Effect | p | Effect | p | Effect | p | Effect | p |
| 1350–1420 | -1.96 | .02 | -2.15 | .11 | -3.47 | .001 | -5.93 | .002 | -4.67 | .001 |
| 1420–1500 | -1.36 | .11 | – | – | – | – | -5.52 | .001 | -3.72 | .001 |

Table 2.11. *Regression analysis identifying constraints on competition between* ne...not *and* not.

| Period | Subordinate clauses | | Neg Conj | | Conditional clauses | | Counterfactual *If*-clauses | | Dependent on neg-clause | |
|---|---|---|---|---|---|---|---|---|---|---|
| | Effect | p | Effect | p | Effect | p | Effect | p | Effect | p |
| 1250–1350 | 0.16 | .64 | -0.31 | .63 | – | – | – | – | – | – |
| 1350–1420 | -0.05 | .79 | -0.11 | .80 | 0.86 | .11 | – | – | 0.44 | .37 |
| 1420–1500 | -1.04 | .12 | 1.33 | .19 | 0.13 | .91 | 0.14 | .98 | 0.22 | .85 |

Middle English, in accordance with the Constant Rate Effect.[12] We should effectively regard *ne* > *ne...not* and *ne* > *not* as two contexts in a single change. It is the loss of stage one *ne*, whether in competition with *ne...not* or *not*, that is subject to clause-type constraints in Tables 2.10 and 2.9.[13]

[12] There is some variation in factor weights for conditional clauses and clauses dependent on a negative clause. It is difficult to make much of this, however. These are rather scarce contexts in the data (see Table 2.6 for the numbers), and this may affect the reliability of the factor weights obtained. In main clauses and subordinate clauses, which are numerous, the factor weights effectively line up across the periods, with the exception of 1250–1350 which is anomalous, but comprises very few texts in the PPCME2.

[13] Although it is widely accepted that the Jespersen Cycle involves a change in the status of *not* from adverbial minimiser to negative marker (e.g. van der Auwera (2009)), this change is not marked by a change in the distribution of *not* in Tables 2.9 and 2.10, perhaps because it happens at an earlier stage of the cycle before the ME period, or because it is manifest in ways other than we have examined here. Chapter 4 examines a wider range of syntactic and functional evidence from which we can identify a distinction between adverbial *not* and negative marker *not*.

On the other hand, the competition between *ne...not* and *not* does not differ significantly across clause types in any sub-period. The results of the regression analyses for each sub-period are given in Table 2.11, including the effect sizes for each context. These estimate how the likelihood of finding *ne...not* in each clause type differs from its likelihood of occurrence in main clauses. The p-values associated with each clause type indicate that none of the differences between the contexts reported achieve statistical significance at the $p<.05$ level. In particular, counterfactuals and clauses dependent on a negative clause fall into line with all other clause types in respect of the change from *ne...not* to *not*, even in the last period 1420–1500. These results are consistent with competition between *ne...not* and *not* operating in all clause types both at the same rate and with a single intercept parameter, rather than different intercepts for the different clause types.

Comparison of the results in Tables 2.9–2.11 indicate that there are two types of *ne* in the Jespersen Cycle, each one involved in a different change that (a) behaves differently with respect to grammatical constraints and (b) proceeds at a different rate $ne_1$ competes with not, $ne_2$ co-occurs with not.

(25) a. $ne_1$ > ($ne_2$)...not = (loss of $ne_1$) – distribution varies by clause type (Rate: 2.6 logits/century)

b. $ne_2$ ...not > not = (loss of $ne_2$) – distribution does not vary by clause type (Rate: 7.2 logits/century)

Once they are separated out, the distribution of each type of *ne* is consistent over time, conforming to the Constant Rate Effect. To test this, I perform two regression analyses on the dataset as a whole, examining the interaction of clause type and date of text, one for the loss of $ne_1$ and one for the loss of $ne_2$. After we exclude a single anomalous text *the Ayenbite of Inwit* dated c.1340[14], we find no statistically significant ($p\leq.05$) interactions between clause type and date in either change. This is evidence that both changes progress at a constant rate across all clause types. In the case of $ne_1$, clause type has a significant effect on the intercept parameter. In the case of $ne_2$, it does not. The rates of the two changes also differ. Fitting a logistic curve to each change allows us to estimate the slope of the curve (the rate of change). Loss of $ne_1$ proceeds at a rate of 2.6 logits per century. The loss of $ne_2$ proceeds at a much faster rate of 7.2 logits per century.[15]

Hence in the overall distribution of *ne*, the apparent reweighting of the constraints on *ne* over time seen in Table 2.6 arises because the two types of *ne* in (25) interact and compete in different ways in different clause types. The relationship between *ne* and *not* follows from this analysis. $ne_1$ and *not*

[14] This text represents a different dialect to the others in the corpus – Kentish – and is also a close literal translation of a French text *Somme le Roi*.

[15] This process is necessarily approximate, given the size of the dataset, gaps where particular periods are not represented in the data at all and approximate dates for some of the texts, but it gives a clear indication that loss of $ne_1$ and $ne_2$ proceed differently over time.

are competitors – functional rather than structural competitors – both negative markers. Consequently, the constraints on *not* are the inverse of those on $ne_1$. As it co-occurs with rather than competes with *not*, $ne_2$ is not a negative marker and it is lost independently of competition between $ne_1$ and *not*. Its status is less clear, and this is where I turn my attention next, in Chapter 3.

## 2.4 Conclusions

In this chapter, we have seen that analyses of the Jespersen Cycle differ in how many distinct stages they assume, and how they structure the transitions between them. Quantitative data from diachronic corpora provide evidence for the inventory of forms required in order to account for patterns of variation and change over time. Accounts of the Jespersen Cycle that assume two independent processes of competition, one affecting the realisation of spec,NegP, the other affecting the realisation of $Neg^0$, imply that the distribution of *ne...not* results from the intersection of these independent changes. However, such accounts, for example Frisch (1997) cannot predict or model differences in the distributions of *ne...not* in different grammatical contexts (clause types). Furthermore, these approaches do not structure the constraints on *ne* in a way that is consistent with the Constant Rate Effect, and they result in the distributions of *ne* and *not* being subject to the same constraints, but in opposite ways. This suggests a relationship between the constraints on *ne* and the constraints on *not* which analysing them as the products of two independent changes does not capture.

Instead, I argue that the reason the overall distribution of *ne* does not conform to the Constant Rate Effect is that the overall distribution of *ne* is in fact the product of competition between two types of *ne*, which are subject to different and independent distributional constraints in diachronic corpus data. Once we distinguish these two types of *ne* in the data, the distribution of each type (a) differs and (b) is subject to constraints which are constant throughout Middle English as Kroch (1989) predicts they should be. Of the two types, $ne_1$ competes with *not* and $ne_2$ co-occurs with *not*. This relationship of functional competition between $ne_1$ and *not* explains why their distributions are the inverse of each other with respect to clause-type constraints. It remains to give a more precise formal characterisation of the differences between $ne_1$ and $ne_2$ in Chapter 5 and to explain the nature of the competition between these two in formal linguistic terms; however, we see that $ne_1$ disappears through competition with *not*, while the loss of $ne_2$ requires another explanation.

The failure of the redundant licensing model is problematic for the mechanism of grammatical competition as Kroch (1989) formulates it. Kroch's original formulation of grammatical competition assumes competition between forms that are structurally and functionally equivalent.

Frisch's (1997) model of the Jespersen Cycle is based on these assumptions. He postulates two processes of competition, each affecting the way a particular syntactic position is realised: competition between *ne* and Ø in $Neg^0$ and competition between Ø and *not* in spec,NegP.[16] However, the model I propose dissociates structural and functional competitors. The way the model is structured makes *ne*$_1$ and *ne*$_2$ structural competitors for the $Neg^0$ position, but *ne*$_1$ and *not* functional competitors for marking negation.

The central issue arising from the observations in this chapter is how to characterise the competing elements *ne*$_1$, *ne*$_2$ and *not* in structural and functional terms. As they compete over time, we know that *ne*$_1$ and *not* must be in some sense formally and functionally equivalent negative markers, and *ne*$_2$ different – not itself a negative marker because it co-occurs with *not*. How can we make a formal distinction between *ne*$_1$ and *ne*$_2$ which predicts their distributions and structures the changes involving them as outlined in this chapter? There are several potential ways to distinguish two types of *ne* either semantically or syntactically which make different predictions about the distribution of *ne*$_2$. So what other evidence is there to characterise the distinction between *ne*$_1$ and *ne*$_2$ more precisely?

In Chapter 3, I argue that variation and change in the distribution of *ne* in redundant or expletive negation contexts is a consequence of the Jespersen Cycle as I structure it in this chapter. These data provide further evidence to distinguish *ne*$_1$ and *ne*$_2$, and moreover provide empirical evidence for a particular formal characterisation of *ne*$_2$ in terms of morphosyntactic features. Then, in Chapter 5, I argue for a Minimalist syntactic analysis of the Jespersen Cycle, in which competition is between mutually exclusive morphosyntactic features. This characterises *ne*$_1$ and *ne*$_2$ in a way that accounts for changes in their distribution that in fact go beyond the Jespersen Cycle. This analysis can accommodate the dissociation between structural and functional competitors within the Jespersen Cycle in a syntactically principled way.

Chapters 8 and 9 consider the consequences of making a syntactic distinction between *ne*$_1$ and *ne*$_2$ in other contexts where *ne* appears, such as negative concord. I argue that the distinction between *ne*$_1$ and *ne*$_2$ provides a new way to make sense of the links between the Jespersen Cycle, changes in negative inversion and the loss of negative concord.

16 Perhaps the greatest difficulty for a model which assumes the independence of *ne* and *not* is having to explain why the supposedly independent changes affecting $Neg^0$ and spec,NegP are subject to the same clause-type constraints, and appear to be the inverse of each other.

# 3 Distributional Evidence for Two Types of *ne*

*Redundant Negation*

## 3.1 Introduction

Chapter 2 argued for a distinction between two types of *ne* – $ne_1$ and $ne_2$ – on the basis of patterns of variation and change in diachronic corpus data within a model of change as morphosyntactic competition. These two types of *ne* have distinct distributions in corpus data and exist in different relationships to *not*: $ne_1$ competes with *not*, and $ne_2$ co-occurs with *not*. This chapter will observe a similar distinction between $ne_1$ and $ne_2$ in contexts of redundant or paratactic negation where *ne* is used independently of *not*. Redundant *ne* provides further evidence of the distinction made between $ne_1$ and $ne_2$ in the preceding chapter. Here, I argue that this evidence allows us to characterise more precisely the distribution – and therefore the syntax – of *ne* at each stage of the Jespersen Cycle.

The phenomenon of redundant negation, also called expletive negation (van der Wurff, 1999b) or paratactic negation (Jespersen, 1917) is well known and extensively described for European languages, see for example van der Wouden (1994), Espinal (2000). Van der Wouden (1994, 107) defines paratactic negation as follows:

> ...various languages and dialects show the effect that a verb (or something else) of negative import triggers a superfluous negation in a subordinate clause ... (van der Wouden, 1994, 107)

An example is given in (26), in which paratactic negation is indicated by PN.

(26) She silly Queene forbad the boy he should not passe those grounds
She silly queen forbade the boy he should PN pass those grounds
'She, silly Queen, forbade the boy to pass those grounds'
(1599 Shaks, Pass Pilgr. 124)

While the typical contexts for paratactic negation (PN) are well known crosslinguistically, data from two early English corpora – the York Corpus of Old English (Taylor et al., 2002) and the PPCME2 (Kroch and Taylor, 2000) – reveal contexts for PN in Middle English which are not described in the literature. They do not involve the usual triggers for PN: predicates of prohibition, fear or denial, comparatives of inequality or prepositions indicating temporal sequence. Hence they are difficult to accommodate under accounts of PN based on negative polarity item licensing (van der Wouden, 1994); on the syntactic subcategorisation of PN contexts, for example van der Wurff (1999b); or on a characterisation of PN contexts in semantic terms (Yoon, 2011).

Instead, there are two types of early English PN, which are distinguished both by properties of the licensing context, and by the negative markers which appear in each type. The licensing contexts for redundant *ne* shift over the course of Middle English, in step with the transition from $ne_1$ to $ne_2$ at stage two of the Jespersen Cycle. Their distribution in PN contexts provides empirical basis on which to characterise $ne_1$ and $ne_2$ in semantic and syntactic terms, in a way that links changes in PN to the Jespersen Cycle.

## 3.2 Paratactic Negation in Early English

### *3.2.1 The Contexts for PN in Early English*

Early English exhibits PN in all the contexts for PN which are most often observed and discussed in the literature, as shown in the following examples. van der Wouden (1994) and Espinal (2000) observe that contexts for PN in a range of languages include the complement clauses of predicates of prohibition (27), fear (28) or denial (29), and comparatives of inequality (30). Early English PN thus follows the pattern established in other languages such as French, Spanish and Dutch. I term this type of paratactic negation PN1 to distinguish it from the second type discussed below.

(27) he deffendeth that man sholde **nat** yeven to his broother ne to his
he forbids that man should PN give to his brother nor to his
freend the myght of his body
friend the might of his body
'he forbids man to give his brother or friend power over his body'
(Chaucer, Melibee 1756)

(28) ðeah for eaðmodnesse wandiað ðæt hi hit **ne** sprecað
yet for humility fear that they it PN preach
'yet for humility they fear to preach them'
(Cura Pastoralis 117.13)

(29) ða wiðsoc Crist swiðe rihtlice þæt he deofol on him **næfde**
then denied Christ very truly that he devil in him PN-had
'Then Christ truly denied that the devil was in him'
(Ælfric Catholic Homilies ii. 230.1)

(30) And thanne thilke thing that the blake cloude of errour whilom had
And then such thing that the black cloud of error formerly had
ycovered schal light more clerly than Phebus himself **ne** schyneth
covered shall light more clearly than Phoebus himself PN shines

'And then the same thing that the black cloud of error had covered in the past will light up more clearly than Phoebus himself shines' (Chaucer Boece III M.11,10, van der Wurff (1999b: 299, ex.8))

We also find PN in the complement clauses of some negated verbs in which the verb gains 'a negative import' (to use van der Wouden's (1994, 107) term) under the scope of negation. In (31), *suffer not* 'not allow' is synonymous with 'forbid'.

(31) And ful ententli þat þis portress suffer nat þat þe dore be **nat**
And very earnestly that this porteress allow not that the door be PN
opened but whan þat grete nede askiþ it
opened but when that great need asks it
'And very earnestly this portress does not allow the door to be opened except when great need demands it'
(15th c. Courtest Sustris Menouresses 91.34-35, Ukaji 1999, ex.12)

However, Early English has additional contexts for PN in which PN appears when a complement clause has positive rather than negative import. Van der Wurff (1999b) notes examples of this type, but does not discuss the differences between them and other instances of PN. These contexts include complement clauses of negated verbs like *forleten* 'forsake or neglect' in (32) or *doute* 'doubt' in (33), and noun phrases with negative quantifiers like *nan tweo* 'no doubt'in (34). The import of 'not doubt' is positive – to be certain of – as is the import of 'not neglect' – to ensure or make certain to. As they involve distinct contexts of occurrence, I distinguish two types of PN – PN1 appears in contexts of negative import such as (26)–(31), and PN2 in contexts of positive import, as in examples (32)–(34).

(32) leoue men þenne ȝe gad to scrifte ne forlete ȝe for nane
dear men when you go to confession NEG neglect you for no
scame þet ȝe **ne** seggen þam preoste alle eower sunne
shame that you PN say the-DAT priest all your sins
'dear men, when you go to confession, do not you neglect for fear of shame that you tell the priest all your sins'
(CMLAMBX1,35.455)

(33) no man douteth that he **ne** is strong in whom he seeth strengthe
no man doubts that he PN is strong in whom he sees strength
'No one doubts that that person is strong in whom he can see strength'
(Chaucer Boece II, Pr.6, 93-4)

(34) forþon nis nan tweo þæt he forgifenesse syllan **nelle**
although NEG-is no doubt that he forgiveness grant PN-will
'although there is no doubt he will grant forgiveness'
(Blickling Homilies 65.8)

Examples superficially like PN2 occur in Present-day English data from the British National Corpus (BNC). However, in Present-day English, when *n't* appears in the complement clause of *no doubt*, the complement clause is interpreted as negative, as in (35). Unlike the Middle English examples given in (32)–(34), the lower clause negation is not redundant or paratactic negation in these Present-day English examples.

(35) a. I have no doubt that Rothmans success will not be dented by an advertising ban = I am certain that Rothmans success will not be dented by an advertising ban
b. I have no doubt that she has not been to Archway Tower = I am certain she has not been to Archway Tower

### *3.2.2 The Diachrony of Early English PN*

Van der Wouden (1994) treats markers of PN as a kind of negative polarity item (NPI). Yoon (2011) also distinguishes markers of PN from markers of sentential negation, arguing instead that markers of PN behave like modal markers. However, in early English the markers of PN in complement clauses of verbs of prohibition, fear or denial (the type I labelled PN1) change over time in step with the Jespersen Cycle. They parallel the default marker of sentential negation at each stage. In an Old English example like (36), paratactic negation takes the form *ne*. From the thirteenth century onwards, *not* is used as a marker of PN1, as in (37).

(36) ða wiðsoc Crist swiðe rihtlice þæt the deofol on him **næfde**
then denied Christ very truly that the devil in him PN-had
'Then Christ truly denied that he had the devil in him'
(ÆCHom ii.230.1)

(37) he commandeth and forbedeth faste / Man shal **not** suffre his wife
he commands and forbids firmly / man shall PN allow his wife
go roule aboute
to-go gad about
'he commands and forbids that men should allow their wives to gad about'
(Chaucer Wife of Bath's Prologue 651ff, van der Wurff (1999b, 306))

By contrast, PN2 is marked by *ne* in the complement clauses of the negated predicates *douten* 'doubt', *forletan* 'neglect' even in Late Middle English, as examples (32)–(33) illustrate. In late fourteenth-century and fifteenth-century English, the two types of PN contrast both in the licensing context and in the way PN is marked.[1]

[1] After the loss of *ne*, PN2 all but disappears, with the exception of one example in Queen Elizabeth's sixteenth century translation of Boethius, given as i.

i. nor does not doute that is **not** the feetes office?
nor does not doubt that is **PN** the feet's purpose?
'nor does not doubt that is the feet's purpose?'
(BOETHEL-E2-P2,78.347).

Table 3.1. *The frequency of PN1 following verbs of fear, prohibition and denial in Old and Middle English prose.*

| Period | PN with *ne* | PN w. *ne...not* | PN w. *not* | PN w. another negative[2] | no PN | Total | % PN |
|---|---|---|---|---|---|---|---|
| 850–950 | 11 | – | – | 3 | 3 | 17 | 71.4% |
| 950–1050 | 12 | – | – | 4 | 2 | 18 | 88.9% |
| 1050–1150 | 5 | – | – | 3 | 2 | 10 | 80.0% |
| Total OE | 62.2% (n=28) | – | – | 22.2% (n=10) | 15.6% (n=7) | 45 | 84.4% |
| 1150–1250 | 2 | 1 | – | – | 5 | 8 | 37.5% |
| 1250–1350 | 2 | – | – | – | – | 2 | 100.0% |
| 1350–1420 | – | – | 3 | – | 7 | 10 | 30.0% |
| 1420–1500 | – | – | – | 1 | 9 | 10 | 10.0% |
| Total ME | 13.3% (n=4) | 3.3% (n=1) | 10.0% (n=3) | 3.3% (n=1) | 70.0% (n=21) | 30 | 30.0% |

The diachrony of PN provides a further reason to distinguish two types of PN: the productivity of each type changes independently of the other. The figures in Table 3.1 are based on searches for PN1 contexts within two corpora: the YCOE (Taylor et al., 2002) and the PPCME2 (Kroch and Taylor, 2000). Table 3.1 reports the results of a search for PN in the complements of predicates of prohibition (FORBID, DEFEND, PROHIBIT), fear (FEAR) and denial (WITHSAY, DENY).

PN1 is most frequent in Old English, and its frequency declines in Middle English. Some examples of this type, given as (38), remain in Present-day English BNC data, but they are extremely marginal.

(38) a. Last night on the world in action programme the minister for health, the member for Peterborough, categorically denied that there was no two tier system and he said (pause) I have no evidence of a two tier system developing in the NHS.
b. Neither Morgan nor Engels denied that he had no direct evidence at all for this first primeval stage of total promiscuity

The text from which i. comes is a translation of a French original. Baghdikian (1979) suggests example i. may be the result of a translation error.

2 These contexts include negative quantifiers and negative adverbs used redundantly, as in the following example:

i. Nature deffendeth and forbedeth by right that na man make hym self riche vn to the
Nature defends and forbids by right that PN man make himself rich un to the
harm of another person
harm of another person
'Nature prohibits and forbids by right that any man makes himself right to the harm of another person'
(Chaucer, Melibee, B. 2774)

Table 3.2. *The frequency of PN following negated predicates 'not doubt' and 'not neglect'.*

| Period | PN2 | No PN | Total |
|---|---|---|---|
| 850–950 | 6 | 20 | 26 |
| 950–1050 | 1 | 1 | 2 |
| 1050–1150 | 0 | 1 | 1 |
| Total OE | 24.1% (n=7) | 75.9% (n=22) | 29 |
| 1150–1250 | 3 | 0 | 3 |
| 1250–1350 | 0 | 0 | 0 |
| 1350–1420 | 7 | 1 | 8 |
| 1420–1500 | 1 | 2 | 3 |
| Total ME | 78.6% (n=11) | 21.4% (n=3) | 14 |

Table 3.2 reports the frequency of PN2 following the verbs *douten* 'to doubt' and *forleten* 'to forsake or abandon', as illustrated by examples (32)–(33) given earlier. In contrast to the overall decline in PN1 in Middle English following predicates of prohibition, denial or fear, Table 3.2 shows that PN2, while never particularly common, becomes more frequent in Middle English than in Old English. It is then lost during the fifteenth century.

The YCOE, PPCME2 and PPCEME corpora contain several translations of the Latin text *Boethius's Consolations of Philosophy* – King Ælfred's tenth century translation, Chaucer's c.1380 translation, George Coleville's translation from 1556, Queen Elizabeth's translation from 1593, and Richard Preston's 1695 translation. Comparison of these different translations of the same text illustrates the diachrony of PN2, and indicates a change in the distribution of *ne* in Late Middle English. In the complement clauses of negated verbs of dubitation (Old English *tweogan* and Middle/Early Modern English *doubt*), only Chaucer's version employs PN2 categorically (n=5/5). There are 10 complement clauses of the verb *ne tweogan* 'not doubt' in King Alfred's translation. Of these, only 2 involve clear examples of *ne* in PN2. These are given in (39).[3]

[3] The interpretation of a third example is not clear – though a paratactic reading of the lower clause negation (*næbben*) seems more appropriate here 'No wise man need doubt that the evil have eternal reward...' than a truth-conditional reading 'No wise man need doubt that the evil do not have eternal reward...'

i. Forðæm ne þearf nænne wisne mon tweogan þæt ða yflan næbben eac ecu
Therefore NEG need no wise man doubt that the evil PN-have each eternal
edlean hiora yfles;
reward of-their evils
'Therefore no wise man need doubt that the evil each have eternal reward for their evils'
(Ælfred, tenth century, coboeth,Bo:37.113.21.2245)

(39) a. Ne þonne ma gif he hwæt bið, ne tweoð nænne mon þæt he
Nor then more, if he that is, NEG doubts no man that he
hwæt ne sie.
that PN is
'Nevertheless, if he is, no man doubts that he is'
(Ælfred, tenth century, coboeth,Bo:16.38.2.687)

b. Ne tweoð nænne mon, gif he wite hæfð, þæt he næbbe yfel.
NEG doubts no man, if he wit has, that he PN-has evil
'No man, if he is in possession of his wits, doubts that he has evil in him'
(Ælfred, tenth century, coboeth,Bo:37.113.29.2250)

Some examples where Chaucer uses expletive *ne* (PN2), but Ælfred lacks it are given in (40) and (41).

(40) a. Ne mæg nænne mon þæs tweogan þætte anwald & genyht
NEG can no man this doubt that power and abundance
is to weorðianne.
is to be reverenced
'No man can doubt that this power and abundance is to be reverenced'
(Ælfred, tenth century, coboeth, Bo:33.75.13.1400)

b. "Certes," quod I, "it nys no doute that it nys right worthy to
"Truly," said I, "it NEG-is no doubt that is PN-is right worthy to
ben reverenced."
be revered"
' "Truly," I said "there is no doubt that it is worthy to be worshiped"'
(Chaucer, c.1380, CMBOETH,429.C1.25)

(41) a. ðu noht ne tweoge þætte God þisse worulde rihtwisige,
you not NEG doubt that God this world directs aright
'Do not doubt that God directs this world righteously and properly'
(Ælfred, tenth century, coboeth,Bo:5.12.26.176)

b. And of alle thise thinges ther nis no doute that thei ne ben
And of all these things there NEG-is no doubt that they PN are
doon ryghtfully and ordeynly...
done righteously and orderly...
'and of all these things there is no doubt that they are done righteously and properly'
(Chaucer, c.1380, CMBOETH,453.C2.536)

c. But thou, allthough thou knowest not the cause of so greate an
But you, although you know not the cause of so great an
order ...doubte thou not all thinges rightly orderd be.
order ...doubt you not all things rightly ordered are

'But you, although you do not know the cause of so great an order ...do not doubt that all things are rightly ordered'
(Queen Elizabeth, 1593, BOETHEL-E2-P2,90.477)

Queen Elizabeth's translation from 1593 contains the same example (41c), but lacks PN in the complement clause, as (41c) shows. Early Modern English *not* does not take on the marking of PN in this context once *ne* has been lost. In all the Early Modern English versions, which postdate the loss of *ne*, there is only one example of PN2 marked by *not* (42).

(42) nor does not doute that is nòt the feetes office?
nor does not doubt that is PN the feet's purpose?
'nor does not doubt that is the feet's purpose?'
(Queen Elizabeth, 1593, BOETHEL-E2-P2,78.347).

Coleville's earlier sixteen-century translation (43) and Preston's later seventeenth-century translation (44) of the same sentence both lack *not* in the complement clause.

(43) for noo man doughteth but that he that maye go by naturall offyce of
for no man doubts but that he that may go by natural office of
hys fete
his feet
'for no man doubts that he may go by the natural function of his feet'
(George Coleville, 1556, BOETHCO-E1-P2,92.347)

(44) And thou doubtest not, but to perform this Motion is the natural
And you doubt not, but to perform this motion is the natural
Office of the Feet?
office of the feet?
'And you don't doubt that to perform this motion is the natural function of the feet?'
(Richard Preston, 1695, BOETHPR-E3-P2,165.378)

This is evidence against a possible counter-argument – that the use of PN2 in Chaucer's *Boethius* may be due to the influence of Latin or French versions of the text upon his translation. First, PN2 is not restricted to Chaucer's translation of the *Boethius*, it occurs in other texts that are not translated from Latin or French. Second, that Chaucer uses PN2 consistently in all contexts where it can appear suggests that its appearance is not accidental but a conscious and systematic choice. Third, PN2 (with *ne*) is systematically distinguished from the sentential negative markers *ne...not* and *not* and from the marker of PN1 *not*. Fourth, even if Chaucer's PN2 is the result of his translation practices, the contrast between Chaucer's translation and the earlier Old English and later sixteenth- and seventeenth-century translations suggests that there is something particular about fourteenth-century English *ne* which allows PN2 to appear in the contexts where it does, and which

Table 3.3. *The properties of two types of paratactic negation.*

| | PN1 | PN2 |
|---|---|---|
| Licensing verbs | Fear, prohibit, deny, not allow, not desire, not be certain | Not doubt, not neglect |
| Context for sub clause | Negative import | Positive import (but only within the scope of negation) |
| Forms used in PN | Sentential negative marker (Old English *ne*, Middle English *not*) | *ne* (at a time when the default marker of sentential negation is *not*) |
| Period of greatest productivity | Old English (9th–11th centuries), although it remains possible in later periods, including Present-day English | Middle English (13th–15th centuries). It is absent from later periods. |

distinguishes it from Old English *ne* and Middle English or Early Modern English *not*. The grammar of Middle English *ne* must be such as to allow it to be used to mark PN2. It is thereby distinguished from Old English *ne* and Early Modern English *not*, which are not used to mark PN2 in earlier and later translations of the same source.

### *3.2.3 Summary*

A semantic and syntactic account of PN needs to take account of two different patterns of PN that differ in their contexts of occurrence, their form and their periods of highest productivity. Table 3.3 summarises the differences. A formal analysis should predict these distributional differences, and the parallels between PN1 and sentential negation strategies. In Section 3.3, I argue that PN1 and PN2 need to be analysed separately. An account that attempts a unified explanation of PN cannot explain this disjunct distribution, or the diachronic changes involved in PN.

## 3.3 Accounting for Early English PN

This section examines three approaches to the ME data presented in Section 3.2: PN as a subtype of NPI licensing (van der Wouden, 1994), PN as paratactic negation (van der Wurff, 1999b), and PN as evaluative negation (Yoon, 2011). Each account makes different predictions concerning the distribution of PN.

### *3.3.1 Parallels between PN and NPI Licensing?*

Both van der Wouden (1994) and Espinal (2000) argue the key property of PN is that it occurs in non-veridical contexts. Van der Wouden (1994) proposes that contexts for PN are monotone decreasing, a property shared with the contexts for NPIs, as (45) illustrates.

(45) I forbid anyone to do that ⟹ I forbid John to do that

While it seems attractive to postulate a unified semantically based analysis of PN in these terms, this approach faces four problems. The first problem is that an account of PN as a subtype of NPI licensing makes a distinction between markers of sentential negation as negative markers and markers of PN as NPIs, implying that different negative forms will be used to mark PN. This is not the case in PN1 where PN typically involves the default marker of sentential negation at each stage of the Jespersen Cycle (*ne* in Old English, *not* in later Middle English). Treating markers of PN as NPIs does not explain these parallels between PN marking and sentential negation marking at successive stages of the Jespersen Cycle.

Second, it makes predictions about the contexts for early English PN which are not borne out. As van der van der Wurff (1999b, 297) observes, it predicts PN will be possible in the complement clause of any negated predicate, since the negative operator creates a monotone decreasing context. However, PN is restricted to the complement clauses of particular subsets of negated predicates.

Third, there are some monotone decreasing contexts in which PN is not found. For example, in the affirmative (46), DOUBT is a monotone decreasing context – (46a) entails (46b) – yet we do not find examples of PN like (46c) in this context.

(46) a. I doubt that anyone will come ⟹
b. I doubt that John will come
c. I doubt that John will not-PN1 come (unattested)

While the property of being a monotone decreasing context may be a necessary condition for PN, it is clearly not a sufficient condition – not all monotone decreasing contexts license PN. We only find examples of PN in the complement of DOUBT when DOUBT is negated, an example like (46c) is unattested in the Old English or Middle English data, either with the Old English form *tweogan* 'doubt' or the Middle English form *doute* 'doubt' (from medieval French).

Fourth, PN2 is not found in monotone decreasing contexts. Example (46) demonstrates that English verbs like DOUBT are monotone decreasing contexts. The proposition 'I doubt that anyone will come' (46a) entails 'I doubt that John will come' (46b). However, when negated, DOUBT ceases to introduce a monotone decreasing context. The proposition (47a) does not entail (47b). Further evidence of this is that a positive polarity item like *someone* seems much more idiomatic here than a negative polarity item *anyone*.[4] This indicates a negated verb of DOUBT is a monotone increasing

[4] At least for the author, who is a native speaker of a British English dialect. To me at least, there is a clear contrast between (47a) with a negative polarity item and (47d) with a positive polarity item, consistent with the polarity reversal in the licensing context. However, analysing PN2

context – (47c) entails (47d) – yet this is exactly the context for PN2 in late Middle English.

(47) a. I don't doubt that John met someone/?anyone $\not\Rightarrow$
b. I don't doubt that John met Mary
c. I don't doubt that John met Mary $\Longrightarrow$
d. I don't doubt that John met someone

Unifying all contexts for PN under a single semantically based explanation such as downward monotonicity is insufficiently restrictive to explain the distribution of PN1 – there are contexts of downward monotonicity where PN1 does not appear. It also does not predict the occurrence of PN2, or allow us to formalise the distinction between PN1 and PN2 in terms of the semantics of their contexts of occurrence. An account based on downward monotonicity predicts that DOUBT will only license PN when it is not negated, the opposite of what we in fact find. In contrast, examples of PN2 in early English appear in monotone increasing rather than monotone decreasing contexts.

### *3.3.2 PN Triggered by Properties of the Context*

Van der Wurff (1999b) and Zeijlstra (2004) suggest that PN is triggered by properties of the context in which it occurs. These approaches face the problems of how to characterise the contexts or predicates which allow PN, how to constrain its distribution – on what basis to distinguish contexts that allow and do not allow PN – and how to explain its optionality.

Looking specifically at early English PN, van der Wurff (1999b) distinguishes predicates like HINDER, FEAR or DENY as introducers of PN because they entail negative embedded propositions. HINDER can be paraphrased as CAUSE P NOT TO BE THE CASE, FEAR as BELIEVE/HOPE THAT P IS NOT THE CASE and DENY as SAY THAT P IS NOT THE CASE. Predicates like HATE or DISLIKE do not license PN because these do not entail a negative embedded proposition. They are paraphrased as NOT LIKE THAT P, with negation taking scope over the matrix predicate rather than just its complement clause.

While this gets us a more restricted distribution for PN than accounts based solely on downward entailment, these kinds of approaches raise two problems. First, PN is optional, even in the complements of predicates that license it. This suggests that there may be more restrictive or specific conditions on its use than accounts like van der Wurff's describe. Second, van der Wurff's V THAT NOT X schema does not accommodate the second type of PN I identified in Section 3.2. The predicates *doubten* 'doubt' or *forleten* 'neglect' do not entail negative embedded propositions, for example DOUBT

*ne* as a positive polarity item also overgeneralises its distribution, predicting its occurrence in contexts where it is systematically absent.

entails NOT BE CERTAIN THAT X not BE CERTAIN THAT NOT X, and NEGLECT entails NOT ENSURE THAT X rather than ENSURE THAT NOT X. Van der Wurff's approach predicts that they do not license PN. His account accommodates PN1, but does not explain why PN2 appears under a certain set of negated verbs whose complement clauses are contexts of positive import, or why PN1 and PN2 take different forms. This suggests that PN2 requires a distinct analysis.

### *3.3.3 Evaluative Negation (Yoon 2011)*

Yoon (2011) argues that PN appears in non-assertive contexts. In these non-assertive contexts, negation does not assert the logical falsity of a proposition – it is not interpreted truth conditionally or in a semantically compositional way. These contexts are non-veridical. Under her account, PN functions to mark the subject's attitude towards or evaluation of a proposition. She identifies two dimensions along which a proposition can be evaluated: epistemic (likelihood or unlikelihood) and bouletic (desirability or undesirability). Yoon argues that PN marks a negative evaluation on one or other of these dimensions: she claims that PN in European languages is primarily bouletic, indicating the desirability of something not happening. It may also be epistemic expressing the likelihood of something not happening.[5]

Yoon's (2011) hypothesis seems a good fit to some of the early English PN data. Thus following verbs of prohibition or denial, PN reinforces the (claimed or perceived) unlikelihood of the state of affairs described by the complement clause, as example (48) shows.

(48) You may deny that you were not the meane of my Lord Hastings late imprisonment
You may deny that you were PN the cause of my Lord Hastings late imprisonment
'You may deny you were the cause of my Lord Hastings's recent imprisonment'
= it is likely that you were not the cause of Lord Hastings's recent imprisonment
(Richard 3, I.iii.502–503, van der Wurff 1999b: 301, ex.14)[6]

In (49), PN evaluates the complement clause negatively, as a strongly undesirable course of action. Similarly, (50) could also be analysed in bouletic terms – the complement clause describes a course of action that is undesirable.

5 I would add a deontic dimension too on the basis of the English data discussed below – the obligation not to do something. Some instances of PN1 seem to indicate a strong obligation against a particular course of action, as in example (49).

6 The unlikelihood of the complement clause proposition being true, or the likelihood of it being false is already established by the preceding context, in which Queen Elizabeth asserts that she was not the cause of Lord Hastings's imprisonment.

(49) he deffendeth that man sholde nat yeven to his broother ne to his
he forbids that man should PN give to his brother nor to his
freend the myght of his body
friend the might of his body
'he forbids man to give his brother or friend power over his body'
= it is desirable that you do not give your brother or friend power over your body
(Chaucer, Melibee 1756)

(50) ðeah for eaðmodnesse wandiað ðæt hi hit ne sprecað
yet for humility fear that they it PN preach
'yet for humility they fear to preach them'
= It is desirable that they do not preach them
(Cura Pastoralis 117.13)

There are some Middle English examples also with *ne*, two of which are given in (51).

(51) a. For certes, ther bihoveth greet corage agains Accidie, lest that
For truly, there needs great courage against Accidie, lest that
it ne swolwe the soule by the synne of sorwe, or destroye it by
it PN swallows the soul by the sin of sorrow, or destroy it by
wanhope.
false-hope.
'Certainly one needs great courage against sloth, lest it swallows the soul by the sin of sorrow or destroy it by false hope'
= it is desirable that sloth does not swallow the soul…
(CMCTPARS,313.C1.1057)

b. "Denyest tow," quod sche, "that alle schrewes ne ben worthy to
"Deny you," said she, "that all sinners PN are worthy to
han torment?"
have torment?"
' "Do you deny," she said "that all sinners are worthy to suffer torment?" '
= is it likely/desirable that sinners do not deserve to suffer torment?
(CMBOETH,448.C2.421)

In (51a), the subordinate clause is an irrealis context, describing a hypothetical consequence of a situation – a consequence which is evaluated as highly undesirable in the context. In (51b), the subordinate clause is again non-assertive, but asserts a proposition that is presupposed to be unlikely in the context, hence the negative evaluation marked by PN. We find similar uses of PN with *ne* in *before* adverbials, as in (52). These also look like evaluative negation, in that the proposition expressed by the adverbial clause describes a situation in both (52a) and (52b) that is evaluated as undesirable.

(52) a. Kyng Arthure hade duellede in þe contre but a litil while, þat
King Arthur had dwelled in the country but a little while, that
men ne tolde him þat þer was comen a grete Geaunt into
men PN told him that there was come a great Gaunt into
Spayne, and hade rauesshede faire Elyne, þat was cosyn to Hoel
Spain and had ravished fair Elaine, that was cousin to Hoel
of Britaigne
of Britain
'King Arthur had dwelled in that country only a little while before men told him that a great Gaunt had come to Spain and had ravished fair Elaine, cousin of Hoel of Britain.'
(CMBRUT3,84.2538)

b. And noþelesse þe trewes durede nouȝt longe, þat þe Danois
And nevertheless the truce held not long, that the Danes
ne bigonne strongeliche forto were oppon þe Englisshe-men
PN began strongly to war upon the English-men
'And nevertheless, the truce did not hold for long, before the Danes began to fight against the English men'
(CMBRUT3,111.3379)

(53) And thanne thilke thing that the blake cloude of errour whilom had
And then such thing that the black cloud of error formerly had
ycovered schal light more clerly than Phebus himself ne schyneth
covered shall light more clearly than Phoebus himself PN shines
'And then the same thing that the black cloud of error had covered in the past will light up more clearly than Phoebus himself shines'
(Chaucer Boece III M.11,10, van der Wurff (1999b: 299, ex.8))

In a comparative of inequality, such as (53), negation is implicit in the comparison. Rephrasing the comparative makes the negation explicit: 'Phoebus [the sun] shall not shine more clearly than the thing that is illuminated'. According to Yoon (2011, 52), PN in comparatives introduces an unlikelihood scale. Thus PN expresses the evaluation that it is likely that the sun will not shine more brightly than the thing which becomes illuminated in (53).

Despite its application to historical English data, there are problems with Yoon's account. Yoon (2011) observes parallels between PN and subjunctive mood. She argues that PN is distinct from sentential negation, being both a marker of negation and of non-assertive mood, and hence restricted to non-veridical contexts which have an evaluative interpretation. In languages like French or Italian, in which markers of PN and markers of sentential negation are different in form, we can follow Yoon's analysis. She argues that markers of PN are similar in function to markers of subjunctive mood in introducing an evaluative (bouletic or epistemic) reading to the clause. However, PN is specifically used to introduce a negative evaluation, so is both negative and epistemic or negative and bouletic in interpretation.

In English evaluative negation, however, there is no distinct form for marking PN. Section 3.2.2 showed that the default marker of sentential negation also functions as a marker of PN1. During the history of English, the marker of PN following verbs of prohibition, fear or denial changes in line with each stage of the Jespersen Cycle, from Old English *ne* to Middle English *not*. If PN1 simply involves a sentential negative marker, it follows that the element which licenses the evaluative reading of PN1 is separate from the marker of negation itself. Two inter-related issues arise here: first, why does the negative marker not contribute negative force to the clause in terms of its truth conditions; and second, if the marker of PN is not the source of the evaluative reading itself, how does such a reading arise?

Turning first to the question of negative force in the subordinate clause, non-assertive subordinate clauses differ from assertive ones. Giorgi (2004) argues that non-assertive embedded clauses lack a force projection or force feature in CP, with the result that these clauses are non-veridical and underspecified for force. She observes on the basis of Present-day Italian data that non-assertive clauses (marked by the subjunctive mood in Italian) have a special status with respect to syntactic locality. She further observes that subjunctive clauses do not block movement out of the clause in the same way that indicatives do, and also that agreement relations can be established between anaphors in subjunctive clauses and their main clause antecedents, in a way that is not possible when the anaphor is in an indicative clause. Giorgi argues that these properties follow if subjunctive clauses lack their own illocutionary force. They are syntactically deficient, lacking a force feature on $C^{0}$ (or a force projection in an articulated CP structure), and are underspecified for force. It follows that negation in these non-assertive clauses will not be interpreted in terms of force, but in some other way.

In order to operationalise Yoon's analysis syntactically in a way that separates the evaluative (modal) and negative elements in the subordinate clause, I hypothesise that a sentential negative marker is interpreted within the scope of deontic, bouletic or epistemic modality, introduced by the main clause verb as it selects a non-assertive clausal complement. Given this, we can adopt a variant of van der Wurff's (1999) analysis (discussed in Section 3.3.2) where there is an implicit (evaluative) negative in the embedded clause, which may be made explicit by the use of PN1 and interpreted relative to the main clause verb. For example FORBID = 'oblige that not P' (interpreted in the scope of a deontic operator), DENY = 'believe that not P' (interpreted in the scope of an epistemic operator), FEAR = 'want/desire that not P' (interpreted in the scope of a bouletic operator).

However, Yoon's analysis goes further than van der Wurff's. Yoon hypothesises that the interpretation of negation in these non-assertive contexts is scalar, indicating speaker evaluation or judgement on a particular modal dimension, for example in the case of FORBID a high degree of obligation against pursuing a particular course of action, in the case of DENY a strong belief that not P, and in the case of FEAR a strong aversion to P or expression of P's undesirability. The optional use of PN1 will therefore vary dependent on the speaker's or writer's subjective evaluation of the proposition in one of these modal dimensions. This is Yoon's central claim, hence the term 'evaluative negation'. A speaker or writer's belief in the likelihood of a particular proposition will also be affected by the context, whether or not the proposition is invoked in the preceding discourse or is presupposed in its negative or affirmative form. These factors will affect any assessment of its epistemic likelihood. In this way, we can maintain that PN1 is both semantically contentful (negative evaluation), but also optional, and hence variable.

Some examples of PN1 in the complements of negated predicates may also be accounted for under this hypothesis, for example, the two examples given by Ukaji of wide scope negative concord with *not*, repeated here as (54) and (55).

(54) And ful ententli þat þis portress suffer nat þat þe dore be
And very earnestly that this porteress allows not that the door be-SBJ
nat opened but whan þat grete nede askiþ it
PN opened but when that great need asks it
'And very earnestly this portress does not allow the door to be opened except when great need demands it'
(15th c. Fifteenth cent. Courtest Sustris Menouresses 91.34-35, Ukaji 1999, ex.12)

(55) I weynd nott þat he scholde not have leyvd tyll þe morning, in so
I believed not that he should PN have lived till the morning, in so
moche þat by my trowthe I dare seye þat iff it had nott fortunyd vs
much that by my honour I dare say that if it had not fortuned us
to have comyn to hym, he had not been on lyve on Wednysday
to have come to him, he had not been alive on Wednesday
'I did not believe that he would have lived till the morning, in so much that by my honour I dare say that if we had not been so fortunate as to come upon him he would not have been alive on Wednesday'
(15th c. Paston Letters 301.4-7, Ukaji 1999, ex.18)

In (54), the clausal complement of suffer 'allow' is non-assertive, as we can see from the subjunctive form of the verb BE in the complement clause. This suggests that the negation is not interpreted truth conditionally, but evaluatively. Furthermore, negation of the predicate *suffer not* 'not allow' has the same deontic modality as *forbid* 'oblige that not X', and therefore creates a context for an evaluative negation reading of the subordinate clause negation in the same way as *forbid*. The state of affairs described by the complement clause is evaluated negatively on both bouletic and deontic dimensions. It is both desirable that it does not happen, and there is a strong obligation against it.

Similarly, in (55), the complement clause of *weynd not* 'not believed' explicitly expresses a belief or judgement. The subordinate clause negation expresses an evaluation indicating the strong epistemic unlikelihood of the state of affairs described in the complement clause – i.e. that it was likely that he would not be alive on Wednesday. In this respect *not believe* is similar to *deny* (if one does not believe something, one denies the truth or likelihood of it) and behaves in the same way as *deny* with respect to evaluative negation.

### *3.3.4 The Problem of PN2*

While a modified version of Yoon's (2011) approach can provide a semantic analysis of early English PN1, Middle English PN2 involving *ne* does not fit into this account well. PN2 appears in the complements of the negated predicates of *doubten* 'doubt' and *forleten* 'neglect' in Middle English especially. The complement clauses in (56) express a proposition which is evaluated positively in both bouletic and epistemic dimensions – that is the writer regards it as both strongly desirable and highly likely, or even self-evidently true in the context of the discourse.

(56) a. ther may no man doute that ther nys som blisfulnesse that is
there may no man doubt that there PN-is some happiness that is
sad, stedefast, and parfyt
sure, constant and perfect
'there may no man doubt that there is some happiness that is sure, constant and perfect' = it is likely/certain that there is some constant and perfect happiness
(CMBOETH,432.C1.134)

b. And of alle thise thinges ther nis no doute that thei ne ben
And of all these things there NEG-is no doubt that they PN are
doon ryghtfully and ordeynly, to the profit of hem to whom we
done rightfully and properly, to the profit of them to whom we
seen thise thingis betyde.
see these things happen.
'And of all these things there is no doubt that they are done rightfully and properly, to the profit of those to whom we see these things happen.' = it is likely/certain that these things are done rightfully and properly...
(CMBOETH,453.C2.536)

So the proposition is neither undesirable or unlikely, and PN2 here does not indicate a negative evaluation of the state of affairs described by the complement clause, in epistemic, deontic or bouletic terms – in fact, the evaluation appears to be the opposite of those expressed by PN1. Note that in all such cases, PN2 is expressed by *ne* rather than *not*, so it is not marked by the usual marker of sentential negation, and does not receive a negative interpretation either truth-conditionally or in terms of evaluation.

Instead of having an evaluative interpretation, PN2 simply looks like wide scope negative concord between a negative marker in a superordinate clause and a concordant negative word in a non-assertive subordinate clause. It is the superordinate clause negation that is semantically interpreted. The subordinate clause negation is not. Giorgi's (2004) analysis in which non-assertive clauses are underspecified for force allows wide scope negative concord across a non-assertive clause boundary, since there is no intervening force feature to block it. This restricts PN2 to non-assertive subordinate clauses. Furthermore, this approach provides an account of why PN2 involves a different negative item.

Two types of PN are distinguished by their pragmatic function and the form of the PN marker. We can characterise PN1 as evaluative negation. There is a clear contrast between the behaviour of *not* and Late Middle English *ne* with respect to PN. This indicates a syntactic or semantic distinction between Old English *ne* and Middle English *not* as negative markers on the one hand and (late) Middle English *ne* on the other which clearly behaves as something other than a typical negative marker. Furthermore, the distinction appears to be one of semantic interpretation – Old English *ne* and Middle/Modern English *not* are negative markers in both sentential negation (realis contexts) and paratactic negation (irrealis contexts), while Middle English *ne* is not, instead only appearing in negative concord. This parallels the distinction I argued for between the two types of *ne* in Chapter 2. The changing distribution of *ne* in sentential negation and paratactic negation should both result from a single parametric change.

### 3.3.5 *Wide Scope Negative Concord (Ukaji 1999): Evaluative Negation?*

There is one more phenomenon that looks like paratactic negation that we should incorporate into the analysis. Ukaji (1999) observes a pattern in Late Middle English and Early Modern English he terms wide scope negative concord, in which a negative word in a subordinate clause appears to be in negative concord with another negative word in a superordinate clause. He observes this phenomenon with a variety of negative words, such as *no* in (57).

(57) And with-owt he wole restore that, I trowe no man can thynk that
nd without he will restore that, I believe no man can think that
his trete is to no good purpos
his threat is to no-PN good purpose
'And if he will not restore that, I believe no man can think that his threat is to any good purpose'
(1450 Paston Letters 39.51-52, Ukaji (1999, example 13)).

Ukaji (1999, 283, ex.50) characterises the contexts for wide scope negative concord as the finite or non-finite clausal complements of:

**Predicates expressing belief or opinion** as illustrated in (57) and (58): *advise, believe, deem, think, trow, ween*
**Predicates expressing modality** as illustrated in (59): *be pertinent, modal+V, need, will* (main verb)
**Predicates expressing cognition** as illustrated in (60): *know, see* (in the sense of 'understand'), *understand, wit*

(58) I whollde not awyse yow to ressaue none of this mony tyll my
I would not advise you to receive none-PN of this money till my
Loor cum himselfe
Lord comes himself
'I would not advise you to receive any of this money until my Lord comes himself' = it is undesirable to receive this money
(c.1481, Cely Letters 121.13-14, Ukaji (1999, ex.22))

(59) a. Nulle ich þat nan iseo ow bute he habbe of ower
NEG-intend I that none-PN see you but he has of your
meister spetiale leaue
master special leave
'I do not intend that anyone should see you except he who has special leave from your master' = it is undesirable that anyone should see you...
(ca. 1230, Ancrene Wisse 14b. 24-26, Ukaji (1999, ex.8))

b. I wuld not in no wyse ye shuld put your-self in no
I intended not in no way you should put yourself in no-PN
daungere to hym
danger to him
'I did not intend that you should put yourself in any danger from him' = undesirable that you put yourself in danger
(PASTON,I,356.115.3554)

(60) and he told him that he had no wrytynge nor euidens of no swych
and he told him that he had no writing nor evidence of no such
thyng..., ner not wyst were he should haue cnowlage of no
thing..., nor not knew where he should have knowledge of no-PN
swyche thyng
such thing
'and he told him that he had no writing nor evidence of any such thing..., and did not know where he should get knowledge of any such thing'. = unlikely that he could get knowledge of any such thing
(ca. 1459, Paston Letters 152.4-6, Ukaji (1999, ex.24))

However, it is not clear that these really do in fact involve negative concord, or that the negative quantifier in the subordinate clause does not contribute to the semantic interpretation of the sentence. Instead, examples (57)–(60) look like contexts for and examples of evaluative negation. They appear specifically in contexts of negative evaluation. The complement clause in (58) is evaluated negatively in bouletic terms (it is desirable that you do not receive the money). The bouletic force of 'not advise' in (58) is not as strong as predicates such as FORBID or PROHIBIT, but it nonetheless involves a reading of undesirability that triggers evaluative negation in the subordinate clause.

Similarly, the complement clauses in (59) are evaluated negatively on a bouletic dimension (it is desirable that you are not seen except...). Finally, in (60) the complement clause is evaluated negatively in terms of likelihood (it is likely he will get no knowledge of the thing) as *not wyst* introduces an epistemic reading of uncertainty.

These data indicate, unsurprisingly perhaps, that Middle English and Early Modern English negative quantifiers are negative markers rather than concordant negative items.[7] They can receive a negative evaluation reading in the same range of non-assertive contexts in which negative markers receive such a reading. Hence many of the examples identified by Ukaji (1999) are not in fact examples of negative concord across clause boundaries, but examples of evaluative negation in non-assertive clauses.

## 3.4 Conclusion

Analyses of paratactic negation as a unified phenomenon fail to account fully for its distribution in early English. Examination of data from diachronic corpora shows that both the form and the distribution of paratactic negation change during Middle English in ways that existing accounts do not predict. Instead, there are two types of paratactic negation at different points in the history of English. The first type of paratactic negation appears in the clausal complements of verbs of hinderance, prohibition, fear and denial, in temporal adverbials introduced by before and in comparatives of inequality. These employ a semantically contentful negative marker or negative operator in a non-assertive subordinate clause, but one which is interpreted evaluatively within in the scope of a bouletic, deontic or epistemic modal operator. The form of that negative operator changes in parallel with the Jespersen Cycle. This type of PN1 appears to mark an evaluation of a proposition as undesirable or unlikely. In this respect, these early English examples fit Yoon's (2011) analysis of evaluative negation.

By contrast, several observations about the distribution of PN2 and the different meaning associated with it point to a different analysis of PN2. PN2 has four important properties: first, it appears in subordinate clauses which are non-assertive – that is which do not affirm or deny a

[7] That is not to say that there are no examples of negative concord involving *no* in the fifteenth and sixteenth centuries, just that (58)–(60) are not examples of wide scope negative concord.

proposition in a truth-conditional sense; second, the polarity of the immediately superordinate clause is always negative; third, these clauses involve Middle English *ne* rather than Middle English *not*; fourth, the negative marker in the subordinate clause makes no contribution to the semantic interpretation of its clause – either in truth-conditional terms, or in terms of expressing a negative judgement or evaluation – but instead appears to be in concord with the main clause negation in a way that does not affect the interpretation of the complement clause.

The nature of the distinction between PN1 and PN2 is conceptually the same as the distinction between two types of *ne* I proposed in Chapter 2. Old English and Middle English *ne*$_1$ is a negative marker, able to contribute negation (¬) to the clause, whether that negation is interpreted truth conditionally (in veridical contexts) or evaluatively (in non-veridical contexts). However, Middle English *ne*$_2$ is not a negative operator itself, but must appear in concord with a negative operator, either the negative marker *not* or a negative such as the quantifier *no*. We see this in examples like (61).

(61) ther may no man doute that ther nys som blisfulnesse that is
there may no man doubt that there PN-is some happiness that is
sad, stedefast, and parfyt.
constant, unchanging and complete
'No man may doubt that there is some happiness that is constant, unchanging and complete'.
(CMBOETH,432.C1.134)

An approach which distinguishes two forms of *ne*, available at different periods, explains why Middle English *ne* is more often found in PN2 than is Old English *ne*. Evidence from paratactic negation points to a distinction between a negative marker (PN1) and a concordant negative item (PN2). It does not support the analysis proposed by Breitbarth (2009). Her analysis follows Wallage (2005, 2008) in making a distinction between two types of *ne*. However, Breitbarth (2009) argues that verbal proclitic negatives like Middle Dutch *ne/en* become non-veridical NPIs at stage three of the Jespersen Cycle. This analysis overgeneralises the distribution of PN2 for English, predicting Middle English *ne* will appear in non-veridical NPI contexts such as comparative clauses and *before* clauses, when in fact we find Middle English *not* used redundantly in those contexts with an evaluative reading. Instead, what defines the distribution of Middle English PN2 is (wide scope) negative concord, suggesting that Middle English *ne* is an anti-veridical rather than non-veridical NPI. Hoeksma (2014, 65) makes a similar point in relation to Middle Dutch *en/ne*, suggesting that there may be three distinct and competing diachronic stages for Dutch *en/ne* – a negative marker, an anti-veridical NPI in Middle Dutch, and a non-veridical NPI in modern Dutch and Flemish dialects. Middle English exhibits only the first two of these stages, *ne* being lost altogether in the fourteenth to fifteenth centuries before it can become a non-veridical NPI.

# 4 Distributional Evidence for Different Types of *not*

## 4.1 Introduction

Chapters 2 and 3 focused on changes to *ne* at successive stages of the Jespersen Cycle. This chapter focuses on changes affecting *not*. It characterises the grammar and functions of *not* at successive stages of the Jespersen Cycle and integrates them into a model of the Jespersen Cycle as grammatical competition.

The negative marker *not* develops through grammaticalisation of the negative argument *na wiht* 'no creature' (Jespersen, 1917). Many accounts postulate an intermediate adverbial *not* stage within the grammaticalisation process so: negative argument *nawiht* > negative adverb *not* > negative marker *not*, as in (62). What functional and syntactic steps comprise this reanalysis, and how are these manifest in the distribution of *not* in corpus data?

The analysis of bipartite negative forms, such as Middle English *ne...not*, proposed in Chapters 2 and 3 allows for two potential configurations – (62b) in which *ne* is a negative marker and *not* a concordant negative item, and (62c) in which *not* is a negative marker and *ne* is a concordant negative item.

(62) a. Stage one: *ne*$_1$
   b. Stage two: *ne*$_1$ plus a concordant negative item, an adverbial minimiser *not* (*not* derived from *nawiht* 'no creature')
   c. Stage three: bipartite negation *ne*$_2$...*not*
   d. Stage four: *not*

From an empirical standpoint, the issues here are first, what evidence there is to distinguish stage two from stage three of the Jespersen Cycle; and second, how the grammatical properties or functions of *not* change as it becomes a negative marker at stage three. Here, I argue there are two forms of adverbial *not* at stage two whose distributions distinguish them from the sentential negator *not* at stage three: one a focus marking adverb, the other an adverbial minimiser. Both the focus marker *not* and the sentential negator *not* represent reanalyses of the OE minimiser *not*.

These changes in the distribution of *not* correlate with competition between *ne*$_1$ and *ne*$_2$ in exactly the way (62) predicts. Chapters 2 and 3 argued that the changes in the distribution of *ne* are consistent with the loss of its ability to mark sentential scope negation at stage three. The reanalysis of *not* and competition between $ne_1$ and $ne_2$ are interlinked: the reanalysis of *not* brings *not* into competition with $ne_1$, resulting in changes to the distribution of both *ne* and *not* which are entirely predictable if $ne_1$ and *not* are functionally equivalent and $ne_2$ is functionally distinct from both.

## 4.2 The Origins of *not*: An Adverbial Minimiser?

The antecedent of *not* is Old English *nawiht*[1], derived by univerbation of the sequence *na wiht*, literally 'no creature'. A nominal quantificational element, used as an argument (*naht* in (63)) undergoes a syntactic shift to become adverbial (as in (64a) and (64b)).

(63) Nu ge geseoð þæt we æt us sylfum naht nabbað. secgað nu
Now you see that we in our selves nothing NEG-have. Say now
hwæt ge be urum weorcum gesawon
what you of our works saw
' "Now you see that we in ourselves have nothing; say now what you have seen our works." '
(cocathom2,ÆCHom_ II,_ 44:331.144.7453)

Hoeksma (2009) hypothesises two possible lines of development for negative markers like *not* which make different predictions about the distribution of *not* at successive stages of the Jespersen Cycle. In the first, the argument of an ambitransitive verb is reanalysed as a negative marker. This account renders unnecessary stage two of the Jespersen Cycle schematised in (62) – there is no intermediate adverb stage. In the other, the reanalysis proceeds via an intermediate adverb stage – an argument first becomes an adverbial minimiser at stage two. This adverb is subsequently reanalysed as a negative marker at stage three.

These two analyses make different predictions about the number of syntactically distinct forms of *not* in the data. The first predicts one non-argument form of *not* – a negative marker. The second predicts two non-argument forms of *not* which may differ in their distributions. Bayer (2009) observes that English *nothing*, German *nichts* and Dutch *niets*, can all be used as adverbial NPIs. When they are, they typically appear with inherently scalar verbs whose extent or degree can be delimited such as 'care' or 'matter'. Bayer argues that although the restrictor THING loses its theta-role, the restrictor itself remains, and it is the restrictor THING that leads to the NPI interpretation.[2] By extension, once this restriction is lost, the NPI will be reanalysed as

[1] I will refer to OE *nawiht* throughout, but note that various forms are found in OE texts, including *naht, nat, nawt, noht(e), nawuht.*
[2] This argument receives support from PDE examples like *I didn't talk a thing about work* (Bayer 2009, ex.70a) in which *a thing* is an adverbial negative polarity item.

a negative marker. Under this account, *nawiht* starts out quantifying over an argument variable, then as it loses its $\theta$-role *not* is reinterpreted as quantifying over a predicate of inherent scale, measure or extent. Finally, as it is reanalysed as a negative marker, it quantifies over a temporal variable (tense) – with the meaning 'there is/was no time t at which the proposition P is true'. If *not* follows this path, becoming an adverbial minimiser at stage two of the Jespersen Cycle, it will become used as a minimiser with scalar predicates at an intermediate stage in its reanalysis.

### *4.2.1 Predicate-type Constraints*

Non-argument uses of *nawiht* are not very frequent in Old English. In the YCOE corpus , it appears in only 0.7% (n=39/5507) of clauses with the sentential negative marker *ne*. Table 4.1 lists the verbs with which *nawiht* is most frequent in the York-Helsinki Parsed Corpus of Old English Prose (YCOE). It compares the distributions of argument *nawiht* in examples like (63) and non-argument *nawiht* in ones like (64).

(64) a. Ne derode Iobe naht þæs deofles costnung. ac fremode.
NEG hurt Job not the devil's temptation. but profited.
for ðan ðe he wæs fulfremedre on geðincðum and Gode near
because he was perfect in honours and God near
æfter þæs sceoccan ehtnysse
after the devil's persecution
'The devil's temptation did not hurt Job but profited him because he was more perfect in honours and nearer to God after the devil's persecution.'
(cocathom2,ÆCHom_ II,_ 35:262.61.5860)

b. and hit bið swyðe derigendlic þæt hi Drihtnes þeowdom
and it is truly harmful that they God's service
forlætan, and to woruldgewinne bugan, þe him naht to
renounce-SBJ and to earthly-war yield-SBJ though him not to
ne gebyriað
NEG concern
'It is truly harmful that they would renounce God's service and yield to earthly war, though it does not concern them'
(coaelive,ÆLS_ [Maccabees]:827.5370)

The table lists only lexical verbs with which *nawiht* occurs three or more times. The frequency of *nawiht* with each verb is calculated relative to the total number of clauses with *ne*, so only instances of *nawiht* in concord with *ne* are counted. Clear instances where *nawiht* has constituent scope, for example when it modifies a quantifier, an adjective or an adverbial phrase (65) are excluded.

Table 4.1. *The frequency of* nawiht *by verb in the YCOE corpus.*

| Verb | | Argument *nawihit* | | | | Adjunct *nawiht* | | | |
|---|---|---|---|---|---|---|---|---|---|
| | | *nawiht* | *ne* | Total | % *nawiht* | adjunct *nawiht* | *ne* | Total | % adjunct *nawiht* |
| *forstandan* | help, benefit, defend | 2 | 1 | 3 | 66.7% | 1 | 1 | 2 | 50.0% |
| *(ge)tweon* | doubt | 4 | 4 | 8 | 50.0% | 4 | 4 | 8 | 50.0% |
| *(ge)fremman* | profit, do good | 6 | 9 | 15 | 40.0% | 3 | 9 | 12 | 25.0% |
| *gefredan* | feel, perceive, know | 0 | 10 | 10 | – | 4 | 10 | 14 | 28.6% |
| *derian* | injure, hurt, harm | 0 | 25 | 25 | 19.4% | 6 | 25 | 31 | 19.4% |
| *gebyrian* | concern | 1 | 13 | 14 | 7.1% | 2 | 13 | 15 | 13.3% |
| *ondrædan* | fear | 2 | 9 | 11 | 18.1% | 1 | 9 | 10 | 10.0% |
| Scalar predicates | | 15 | 71 | 86 | 17.4% | 21 | 71 | 92 | 29.6% |
| *findan* | find | 2 | 30 | 32 | 6.3% | 1 | 30 | 31 | 3.2% |
| *mæg* | have strength to | 1 | 68 | 69 | 1.4% | 2 | 68 | 70 | 2.9% |
| *gefon* | take, sieze, catch | 3 | 1 | 4 | 75.0% | – | – | – | – |
| *behealdan* | possess, consider, beware, take heed | 3 | 4 | 7 | 42.9% | – | – | – | – |
| *gedon* | make, cause, put | 5 | 18 | 23 | 21.7% | – | – | – | – |
| don | do, make, cause | 12 | 70 | 82 | 14.6% | – | – | – | – |
| *(ge)cwæþan* | say, speak | 3 | 21 | 24 | 12.5% | – | – | – | – |
| habban | have | 6 | 57 | 63 | 9.5% | – | – | – | – |
| *wyrcan* | work | 3 | 29 | 32 | 9.4% | – | – | – | – |
| *sellan* | give | 6 | 59 | 65 | 9.2% | – | – | – | – |
| *geseon* | see | 7 | 90 | 97 | 7.2% | – | – | – | – |
| Non-scalar predicates | | 57 | 447 | 504 | 11.3% | 3 | 98 | 101 | 3.0% |

(65) Nis hit naht feor ascaden from ðisse worulde
NEG-is it not far separated from the world
'It is not far separated from the world'
(cocura,CP:51.399.26.2728)

There is a clear distinction between argument and adverb *not* consistent with its reanalysis as an adverbial minimiser, rather than a negative marker. Adverbial *not* is largely restricted to occur with particular predicates. Although the number of instances is small, the adverb *not* appears most frequently with *forstandan* (66a), *(ge)tweon* (66b), *(ge)fremian* (66c), *(ge)fredan* (66d) and *derian* (66e). Many are psych-verbs, and all express states, experiences or judgements which can be quantified or measured and are consistent with a minimiser interpretation of *nawiht*.[3]

(66) a. Ond suaðeah nu, ðeah se lareow ðis eall smealice &
And nevertheless now, though the teacher this all carefully and
openlice gecyðe, ne forstent hit him noht
openly told, NEG helped it him not
'And nevertheless now, though the teacher told all this carefully and openly, it did not help him'
(cocura,CP:21.163.18.1115)

b. Ne tweoge ic naht, þæt gode weras wæron on þysum lande:
NEG doubt I not, that god men were in this land
'I do not doubt that good men were in this land'
(cogregdC,GDPref_1_[C]:7.12.50)

c. hi habbað swaðeah hoge be heora freondum on life,
They have nevertheless care by their friends on life,
þeah ðe hit naht ne fremige, ne heora freondum ne
though that it not NEG benefit-SBJ, neither their friends nor
him.
them.
'they nevertheless care for their friends in life, though it benefits neither their friends nor them'
(coaelhom,ÆHom_11:268.1630)

d. & ic þine tintregu naht ne gefrede.
and I your torment not NEG feel.
'and I do not feel your torment'
(cocathom1,ÆCHom_I,_29:423.142.5749)

[3] While this is a necessary condition for minimisers like OE *nawiht* or PDE *one bit*, it is insufficient in itself to predict where a minimiser will occur. Where predicates are related by scalar implicatures, these implicatures will ensure that only predicates that are compatible with the minimal point on an extent scale will co-occur with a minimiser. For example in PDE *I didn't love it one bit* is blocked by *I didn't like it one bit*, and *I didn't hate it one bit* blocked by *I didn't mind it one bit*. *Mind* and *like* denote states whose extent or intensity is lower than *hate* and *love*. Minimisers such as PDE *one bit* and OE *nawiht* only readily occur with predicates at the lower points these extent or intensity scales.

e. and þæt fyr ne derede naht þam ðrim cnihtum. ðe on God belyfdon
and that fire NEG harmed not the three youths that in God believed
'and that fire did not harm the three youths who believed in God'
(cocathom2,ÆCHom_I,_ 1:9.241.210)

The adverbialisation of *nawiht* results in a minimiser: the most frequent contexts for adverbial *nawiht* in Table 4.1 are not ambitransitive verbs, but psych-verbs like 'doubt', 'feel' and predicates like 'help' or 'harm' which can be quantified by an adverbial of extent. In this way OE *nawiht* parallels the examples Hoeksma (2009) gives with Present-day English NPI adverbial minimisers *one bit* (67) or *any* (68). This distribution indicates that when the quantifier *nawiht* 'no creature/thing' is reanalysed as an adverbial, it is typically used as a minimiser during Old English.

(67) Jones did not like her one bit (Hoeksma, 2009, 29, ex.22a)
(68) That did not help the soldiers any (Hoeksma, 2009, 29, ex.23a)

The distribution of OE *nawiht* parallels that of PDE *one bit.* Table 4.2 reports the distribution of the NPI minimiser *one bit* used as an adverbial in clauses with *not/n't*, in the British National Corpus (BNC; 100 million words) and the Corpus of Contemporary American English (COCA; 520 million words). Only verbs with which *one bit* occurs twice or more are counted. Although never very frequent with any verb, the verbs with which the OE adjunct minimiser *nawiht* and the PDE adjunct minimiser *one bit* occur fall into two broad groups: psych-verbs that denote emotional states, verbs that express judgements or attitudes that vary according to degree and are therefore measurable (*like, blame, surprise, mind, trust, bother, care, miss, matter, worry, enjoy, believe, appreciate, be/seem happy, feel, scare*) and processes with an effect that can be estimated, measured or quantified by a degree adverbial (*change, hurt, affect, slow down, faze, diminish, improve*).[4] All these predicates are scalar predicates. The distribution of OE *nawiht* parallels that of PDE *one bit.*

[4] Other verbs with which *one bit* occurs once in the BNC are *unwind, impress, appreciate, pine, care, diminish, fool, affect, decrease, relish, alter, agree, dampen, be afraid, overrate, take to, bother, reduce*. In COCA we also find instances of 'one bit' with *doubt, recede, die down, benefit, soften, challenge, deserve, disturb, look safe, think about, suit, take to, waver, be upset, trouble, relish, disappoint, want, be jealous, be sorry, fool, underestimate, regret, subscribe (to an idea), fix, register, back off, vary, compare, dim, tarnish, lessen, darken, back down, damage, drop off, enhance, abate, age, back away, wane, conceal, trivialise, bolster, reduce, deny, give, harm, stop, be overwhelmed, notice, envy, dampen, inhibit*. All these fall within the categories (emotions, attitudes, processes) outlined above.

Table 4.2. *The distribution of the adverbial NPI* one bit *in the BNC and COCA corpora by verb.*

| | BNC | | | COCA | | |
|---|---|---|---|---|---|---|
| Verb | one bit | Total not/n't+V | % one bit | one bit | Total not/n't+V | % one bit |
| like | 18 | 5908 | 0.3% | 111 | 30507 | 0.4% |
| help | 5 | 2431 | 0.2% | 17 | 16443 | 0.1% |
| blame | 3 | 506 | 0.6% | 22 | 3126 | 0.7% |
| surprise | 3 | 164 | 1.8% | 11 | 1336 | 0.8% |
| mind | 2 | 2466 | 0.1% | 19 | 7765 | 0.2% |
| change | 2 | 1292 | 0.2% | 28 | 9930 | 0.3% |
| trust | 2 | 532 | 0.4% | 6 | 3991 | 0.2% |
| bother | – | – | – | 24 | 5509 | 0.4% |
| care | – | – | – | 15 | 16986 | 0.1% |
| hurt | – | – | – | 13 | 5011 | 0.3% |
| miss | – | – | – | 10 | 4232 | 0.2% |
| affect | – | – | – | 9 | 2702 | 0.3% |
| matter | – | – | – | 9 | 11564 | 0.1% |
| worry | – | – | – | 8 | 10404 | 0.1% |
| slow down | – | – | – | 8 | 624 | 1.3% |
| phase/faze | – | – | – | 6 | 278 | 2.2% |
| enjoy | – | – | – | 5 | 1511 | 0.3% |
| believe | – | – | – | 4 | 29578 | 0.01% |
| appreciate | – | – | – | 4 | 1339 | 0.3% |
| be/seem happy | – | – | – | 3 | 2455 | 0.1% |
| feel (with adj) | – | – | – | 3 | 2607 | 0.1% |
| scare | – | – | – | 3 | 512 | 0.6% |
| diminish | – | – | – | 3 | 568 | 0.5% |
| improve | – | – | – | 3 | 1289 | 0.2% |

### *4.2.2 Dating the Reanalysis of* not

Clauses in which *nawiht* is a minimiser provide a context in which it is susceptible to reanalysis. Reanalysis results in *not* generalising from scalar predicates compatible with the minimiser *nawiht* to all verbs. Therefore, the extent to which ME *not* occurs with verbs which are not compatible with the earlier minimiser interpretation of OE *nawiht*, for example with punctual verbs such as *kill* (cf. PDE *He didn't kill her one bit*), provides evidence for the reanalysis of the minimiser *nawiht* as a negative marker. Examining the verbs with which *not* occurs at various points in early ME provides evidence to chart the progress of this reanalysis over time and to incorporate it into a quantitative model of the Jespersen Cycle.

Table 4.3 lists the verbs with which *not* occurs most frequently in the period 1150–1250. Only verbs with which *not* occurs three times or more are listed. Many of the verbs in Table 4.3 are similar to those with which OE *nawiht* appears – scalar predicates, verbs such as *fear* or *help* (cf. Table 4.1) – and consistent with the verbs with which the PDE minimiser *one bit* appears, for

Table 4.3. *The frequency of* not *by verb in the first subperiod of the PPCME2 corpus, 1150–1250.*

| Verb | | *ne...not* | *ne* | Total | % *ne...not* |
|---|---|---|---|---|---|
| *dreden* | fear | 6 | 0 | 6 | 100.0% |
| *lefen* | believe, trust in | 3 | 2 | 5 | 60.0% |
| *lufien* | love | 5 | 4 | 9 | 55.6% |
| *wenen* | believe | 3 | 0 | 3 | 100.0% |
| *speden* | achieve, fulfil | 3 | 0 | 3 | 100.0% |
| *forleten* | forsake, abandon, reject | 8 | 8 | 16 | 50.0% |
| *helpen* | help | 8 | 4 | 12 | 66.7% |
| Scalar predicates | | 36 | 18 | 54 | 66.7% |
| *bringen* | bring | 3 | 0 | 3 | 100.0% |
| *ought* | ought | 3 | 1 | 4 | 75.0% |
| *thorfen* | need | 3 | 1 | 4 | 75.0% |
| *thinken* | think | 9 | 7 | 16 | 56.3% |
| *seon* | see | 5 | 4 | 9 | 55.6% |
| *given* | give | 3 | 3 | 6 | 50.0% |
| *knowen* | know | 3 | 3 | 6 | 50.0% |
| *leten* | grant | 5 | 5 | 10 | 50.0% |
| BE[5] | | 25 | 34 | 59 | 42.4% |
| *cumen* | come | 3 | 5 | 8 | 37.5% |
| *understanden* | understand | 3 | 6 | 9 | 33.3% |
| *holden* | hold | 4 | 9 | 13 | 30.8% |
| *seien* | say | 4 | 9 | 13 | 30.8% |
| DO | | 8 | 19 | 27 | 29.6% |
| HAVE | | 7 | 26 | 33 | 21.2% |
| Other predicates | | 88 | 132 | 220 | 40.0% |
| TOTAL | | 124 | 150 | 274 | 45.3% |

example *believe, help, fear, trust, need, love*. However, ME *not* occurs with a wider range of verbs, many of which (for example, *bring, see, give, come, hold, say, do, have*) are not scalar predicates and do not occur with OE minimiser *nawiht* in the YCOE data, or with the PDE minimiser *one bit* in the BNC or COCA data.

There is clearly a change in the distribution of *not* from OE to early ME, a weakening of the earlier (Old English) link between scalar predicates and *nawiht*/*not*. This change in the distribution of *not* is consistent with reanalysis of *not* from an NPI adverbial minimiser to a negative marker. However, separating the predicates in Table 4.3 into scalar and non-scalar predicates reveals that *not* remains more frequent with scalar predicates like *believe, hope, fear, trust* than non-scalar predicates like *say, come, do, have, give, need, think,*

[5] The status of BE with respect to minimisers such as PDE *one bit* depends on how BE is used. The minimiser *one bit* occurs with BE when BE introduces a scalar adjective such as *happy* (*He wasn't one bit happy about the timing*, Sherryl Woods (2008) *Seaview Inn*, COCA) or *sorry* (*Well, I'm not one bit sorry for you* Alice Taylor (1997) *The Woman of the House*, COCA).

Table 4.4. *The frequency of* not *by verb in the second subperiod of the PPCME2 corpus, 1250–1350.*

| Verb | | *ne...not* | *ne* | Total | % *ne...not* |
|---|---|---|---|---|---|
| *dreden* | fear | 6 | 0 | 6 | 100.0% |
| *douten* | doubt | 3 | 0 | 3 | 100.0% |
| *forsaken* | renounce, neglect | 3 | 0 | 3 | 100.0% |
| *hopen* | hope, believe, expect | 3 | 0 | 3 | 100.0% |
| *tholen* | suffer, submit to judgement | 3 | 0 | 3 | 100.0% |
| *willen* | desire,wish | 5 | 1 | 6 | 83.3% |
| *forgieten* | forget | 18 | 1 | 19 | 94.7% |
| Scalar predicates | | 41 | 2 | 43 | 95.3% |
| *confounden* | defeat, condemn, curse | 9 | 0 | 9 | 100.0% |
| *cumen* | come | 8 | 0 | 8 | 100.0% |
| *hiden* | hide, conceal | 6 | 0 | 6 | 100.0% |
| *thinken* | think | 5 | 0 | 5 | 100.0% |
| *stiren* | set in motion, move something | 5 | 0 | 5 | 100.0% |
| *given* | give | 5 | 0 | 5 | 100.0% |
| *saven* | save | 5 | 0 | 5 | 100.0% |
| *speken* | speak | 5 | 0 | 5 | 100.0% |
| *kepen* | keep | 4 | 0 | 4 | 100.0% |
| *turnen* | turn, roll, twist | 4 | 0 | 4 | 100.0% |
| *bowen* | bow, kneel, be submissive/obedient | 3 | 0 | 3 | 100.0% |
| *singen* | sing | 3 | 0 | 3 | 100.0% |
| *hieren* | hear | 12 | 1 | 13 | 92.3% |
| *setten* | make, sit, move/put in place | 9 | 1 | 10 | 90.0% |
| *sen* | see | 7 | 1 | 8 | 87.5% |
| *understanden* | understand | 6 | 1 | 7 | 85.7% |
| *witen* | know, understand | 5 | 1 | 6 | 83.3% |
| *knowen* | know | 16 | 5 | 21 | 76.2% |
| *sayen* | say | 18 | 6 | 24 | 75.0% |
| *finden* | find | 3 | 1 | 4 | 75.0% |
| *maken* | make | 3 | 1 | 4 | 75.0% |
| *taken* | take | 3 | 1 | 4 | 75.0% |
| BE | | 71 | 24 | 95 | 74.7% |
| DO | | 12 | 5 | 17 | 70.6% |
| GO | | 11 | 6 | 17 | 64.7% |
| *amenden* | remedy, correct,repair | 3 | 2 | 5 | 60.0% |
| *holden* | hold | 7 | 9 | 16 | 43.8% |
| HAVE | | 16 | 24 | 40 | 40.0% |
| Other predicates | | 268 | 89 | 357 | 75.1% |
| TOTAL | | 309 | 91 | 400 | 77.3% |

*understand*. The difference between the frequencies of *not* in these two groups remains significant in the 1150–1250 data: $\chi^2$(1df)=12.144, p<.001.[6]

Table 4.4 lists the verbs with which *not* appears most frequently in the period 1250–1350.[7] Again, many of the verbs which occur most frequently

[6] The distributions of the high frequency verbs BE, HAVE and DO is not responsible for this finding. Excluding these verbs gives a figure for the non-scalar verbs of 47.5% (n=48/101). This remains significantly different from the scalar verbs $\chi^2$(1df)=5.19, p=.022.

[7] Only verbs which occur with *not* more than three times are included.

Table 4.5. *The distribution of* not *by predicate type in the PPCME2 corpus.*

| | 1150–1250 | | | 1250–1350 | | |
|---|---|---|---|---|---|---|
| Predicate | *not* | Total | % *not* | *not* | Total | % *not* |
| Scalar | 65 | 107 | 60.7% | 45 | 49 | 91.8% |
| Non-scalar | 197 | 559 | 35.2% | 436 | 590 | 73.9% |

with *ne...not* in Table 4.4 are scalar predicates, such as *dreden* 'fear', *hopen* 'hope', *tholen* 'suffer', *douten* 'doubt', *forsake* 'renounce/neglect'. However, some of the verbs occurring most frequently with *not* are not scalar predicates, such as *confounden* 'defeat, condemn, curse', *saven* 'save'. Verbs of articulation *speken* 'speak'. *singen* 'sing', verbs of motion *stiren* 'set in motion', *turnen* 'turn', or the senses *sen* 'see', *hieren* 'hear' are also not scalar predicates, but nevertheless frequently occur with *not*. Others, like *given* 'give', *cumen* 'come' were among the least frequently occurring with *not* in the preceding 1150–1250 period, but occur more frequently with *not* in the period 1250–1350. However, BE, DO and HAVE remain resistant to *not* throughout Old English and Early Middle English (to c.1350).

When we categorise all predicates into scalar and non-scalar groups in the Middle English periods 1150-1250 and 1250-1350 (Table 4.5), we see a difference between the distribution of *not* across the two periods. In Middle English, its distribution is no longer exclusively that of a minimiser. The distribution of *not* from Middle English to c.1350 appears to show the gradual generalisation of *not* from scalar predicates to all predicates during the twelfth to fourteenth centuries, with the exceptions of BE, HAVE and DO, which continue to resist *not* until at least the mid fourteenth century.

A regression model provides evidence to date the reanalysis of *not* as a negative marker. When we compare the results of two mixed-effects regression models in Table 4.6, one for the period 1150–1250 and one for the period 1250–1350, we find that the distinction between scalar and non-scalar predicates is consistent across both periods in terms of effect size, although the significance of the effect decreases as *not* becomes more frequent overall.[8] Scalar predicates favour *not* consistently across both periods. That means the rate at which *not* is introduced is the same for both scalar and non-scalar predicates – that the same change is introducing *not* in both contexts equally.

The spread of *not* in Middle English is therefore independent of the distinction between scalar and non-scalar predicates. The only way this independence can arise is if the ME spread of *not* represents ongoing competition between the negative markers *ne* and *not* with all predicates – that is, if *not*

[8] These mixed-effects regression models include the predictors predicate type and clause type and the random effect of corpus text. The reference level for predicate type is non-scalar predicates.

Table 4.6. *Results of mixed-effects regression for predicate type in Early Middle English.*

| Period | Scalar predicates | p |
|---|---|---|
| 1150–1250 | 1.00 | <.001 |
| 1250–1350 | 0.82 | .23 |

is already a negative marker by the twelfth century. That scalar predicates consistently favour *not* simply reflects the higher frequency of *not* in that context at the point prior to ME when the reanalysis takes place.

## 4.3 Negative Inversion and Two Types of *not* in Middle English

### *4.3.1 The Syntax of Negative Inversion in Old English*

This section takes as its starting point the hypothesis that changes in the availability of negative inversion in clauses with *ne...not* distinguish *not* at stage two of the Jespersen Cycle from *not* at stage three. Negative inversion involves V to C movement. Evidence for V to C movement comes from subject-verb inversion, but only with pronominal subjects. Haeberli (2002a) identifies two subject positions in Old English clauses with transitive verbs, spec,TP for nominal subjects and spec,AgrP for pronominal subjects, proposing the clause structure in (69).

(69)

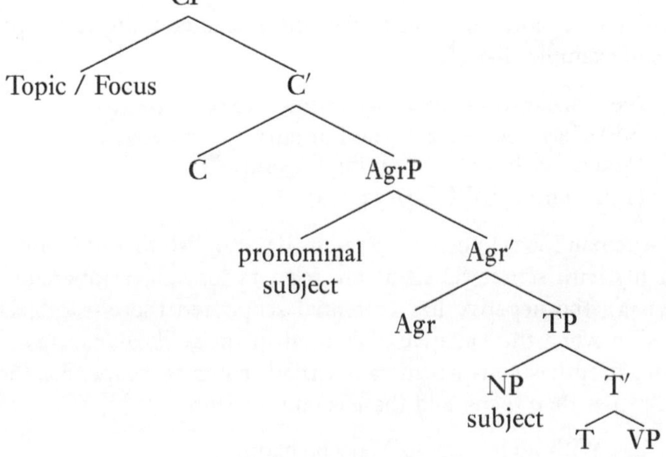


Finite verbs occupy one of three positions relative to these two subject types. Typically in subordinate clauses, finite verbs follow both pronominal and nominal subjects. Haeberli argues this position is $T^0$. However, in main clauses, finite verbs appear in $Agr^0$. Where a non-subject topic stands in spec,CP, the finite verb precedes nominal subjects, but follows pronominal subjects. However, in some clause types, finite verbs move to $C^0$ and can precede both nominal and pronominal subjects. Pintzuk (1999, 90ff) identifies contexts for verb-movement to $C^0$ in Old English including:

(70) Interrogatives:
hwi sceole we oþres mannes niman
why should we another man's take
'Why should we take those of another man?'
(ÆLS 24.188, Pintzuk (1999, 90, ex.111))

(71) Imperative or subjunctive mood:
sie he mid stanum ofworpod
be he with stones slain
'...he should be slain with stones...'
(Laws Af El 21 34.1, Pintzuk (1999, 91, ex.113))

(72) Clauses with initial discourse adverbs eg. *þa* / *þonne* 'then', *nu* 'now':
þa gemette he sceaðan
then met he robbers
'...then he met robbers...'
(ÆLS 31.151, [92, ex.121] Pintzuk (1999))

(73) Negated finite verbs:
Ne forgife ic eow swa swa þes middaneard forgifð
NEG forgive I you as this world forgives
'I do not forgive you as this world does'
(10th century; Ælhom,Æhom_10:15.1413)

Negative inversion – inversion following a negated finite verb – persists into EME in examples like (74).

(74) Ne hafst tu næure soðe eadmodnesse on þe...
NEG have you ever true humility in you...
'You never have true humility in you...'
(13th century; VICES1,33.398)

Haegeman (2001) notes for Present-day English that only initial negative elements with sentential scope are triggers for subject-operator inversion. In (164a), the negative has sentential scope and there is subject-operator inversion when the negative is focused in spec,CP. By contrast, in (164b), the negation has constituent rather than sentential scope, has the status of topic rather than focus, and there is no inversion.

(75) a. With no job would Mary be happy
(=Mary wouldn't be happy with any job)
(Haegeman 2001: 21, ex.1a)

b. With no job, Mary would be happy
(=Mary would be happy without a job)
(Haegeman 2001: 21, ex.1b)

Therefore, in clauses with bipartite *ne...not*, negative inversion provides evidence to the learner that *ne* is the negative operator. This is key to understanding the loss of negative inversion and its relationship to the Jespersen Cycle. If we extend this idea back to Middle English, the two forms of *ne* I propose in Chapters 2 and 3 will differ with respect to negative inversion. $ne_1$ is a negative operator and participates in negative inversion, $ne_2$ is not and does not. It follows that negative inversion and the negative operator *not* are mutually exclusive because both $ne_1$ and *not* are negative operators. However, adverbial *not* and negative inversion can co-occur. The next sections explain how this interaction between negative inversion and the Jespersen Cycle patterns over time, and how it allows us to identify different forms of *not* in inverted and uninverted clauses.

### 4.3.2 *The Loss of Negative Inversion in Middle English*

From the fourteenth century onwards, negative inversion and the negative marker *not* are largely incompatible ways to mark sentential scope negation. In Early Modern English (1500–1710), we find inversion following clause-initial negative items in only 7.7% (n=3/39) of clauses in which the initial negative item co-occurs with *not*, but 94.0% (n=455/484) clauses where the initial negative item does not co-occur with *not*.[9] However, in early Middle English negative inversion and *not* co-occur more frequently in examples like (76) with clause-initial *ne*.

(76) Ne bið naut his lare fremful ne icweme þan ileweden
NEG is not his teaching beneficial nor suitable for the laity
'His teaching is not beneficial nor suitable for the laity'
(CMLAMBX1,109.1006, c.1225)

Although negative inversion and *not* are not mutually exclusive in EME, both Ingham (2005) and Wallage (2012a) observe a link between the loss of negated verb movement to $C^0$ and the spread of bipartite *ne...not* forms. Table 4.7 gives the frequencies of negative inversion in two environments, clauses negated by *ne* alone and clauses negated by bipartite negation (*ne...na* in Old English and *ne...not* in Middle English).[10] The data comprise indicative non-conjoined main clauses with pronominal subjects.[11]

[9] For further discussion of these data on inversion in Late Middle English and Early Modern English see Chapter 8.

[10] The Old English data in Table 4.7 come from the York-Helsinki Parsed Corpus of Old English Prose (Taylor et al., 2002) and Middle English data come from the 2nd edition of the Penn-Helsinki Parsed Corpus of Middle English (Kroch and Taylor, 2000).

[11] Imperatives and subjunctives are contexts for V to C movement irrespective of their polarity, while V to C movement is restricted to indicatives only when they are negated by *ne*. Therefore, in order to isolate the effect of negation from the effect of mood, the data in Table 4.7 are restricted to unambiguously morphologically marked negative indicative verbs only.

Table 4.7. *Negated verb movement to C at successive stages of the Jespersen Cycle.*

| | Unsupported (stage one) *ne* | | | Bipartite (stage 2/3) *ne...na/not* | | |
|---|---|---|---|---|---|---|
| Period | *ne*+V in $C^0$ | Total | % in C | *ne*+V in $C^0$ | Total | % in C |
| 850–950 | 55 | 123 | 44.7% | 33 | 43 | 76.7% |
| 950–1050 | 225 | 440 | 51.1% | 63 | 90 | 70.0% |
| 1050–1150 | 30 | 58 | 51.7% | 16 | 27 | 59.2% |
| 1150–1250 | 4 | 6 | 66.7% | 13 | 34 | 38.2% |

Table 4.8. *Bipartite negation in inverted and uninverted negative clauses. Non-conjoined main clauses with morphologically marked indicative verbs only. Data from the YCOE and PPCME2 corpora.*

| | Neg inverted clauses | | | Uninverted clauses | | |
|---|---|---|---|---|---|---|
| Period | bipartite | Total | % bipartite | bipartite | Total | % bipartite |
| 850–950 | 33 | 88 | 37.5% | 10 | 78 | 12.8% |
| 950–1050 | 63 | 288 | 21.8% | 27 | 242 | 11.2% |
| 1050–1150 | 16 | 46 | 34.8% | 11 | 39 | 28.2% |
| 1150–1250 | 13 | 17 | 76.5% | 21 | 23 | 91.3% |

The frequency of negative inversion is relatively stable in clauses which are negated by *ne* alone at stage one of the Jespersen Cycle. Negative inversion declines from the twelfth to fourteenth centuries, prior to the loss of *ne* itself, but only in clauses with *ne...not*. The trigger for V to C movement in clauses with initial negative elements is polarity focus. Consequently, only the negative operator $ne_1$ triggers V to C movement, the NPI $ne_2$ does not. The negative operator $ne_1$ occurs in negative inversion and co-occurs with the adverb *not*. The NPI $ne_2$ does not participate in negative inversion, as it is not a negative operator. Instead it must co-occur with the negative operator *not*. Competition between the negative operator $ne_1$ and the negative operator *not* results in the loss of negative inversion because they are competing ways to mark clausal negation at LF.[12] In clauses where $ne_1$ remains – those without *not* in Table 4.7 – negative inversion remains productive.

Table 4.8 is based on the same data as Table 4.7, but it reports the frequency of OE *na* and ME *not* in clauses with and without negative inversion. While negative inversion and *not* are mutually exclusive from the 14th century onwards, Table 4.8 indicates that *na* or *not* and negative inversion are not mutually exclusive in OE or early ME. However, ME *ne...not* contrasts with

[12] Competition between $ne_1$ and *not*, the loss of negative inversion are both reflexes of the same underlying change – *not* becoming a negative operator – so both should proceed at the same rate.

OE *ne...na* is more frequent in inverted than uninverted clauses[13], while for EME *ne...not* the opposite is true. This suggests an emergent grammatical distinction between OE *na* and ME *not* consistent with the incipient grammaticalisation of ME *not* as a negative marker.[14]

The loss of negative inversion in clauses with *ne...not* provides evidence for the syntactic reanalysis of *not* from concordant negative item to sentential negative marker. Making the link between *not* and the loss of negative inversion requires competition between two forms of *not*.[15] It does not follow in an account where all instances of *not* in both inverted and uninverted clauses are a single lexical item. In inverted clauses *not* must be a concordant negative adverb, whereas in uninverted clauses, *ne...not* is structurally ambiguous – *not* could be either an adverb or negative marker. Therefore we may see differences in the distributions and functions of *not* in inverted and uninverted clauses. The next section examines evidence for this, addressing first the function of *not* in inverted clauses.

## 4.4 Evidence to Distinguish Two Different Forms of *not* in Inverted and Uninverted Clauses

### *4.4.1* Not *in Negative Inversion*

The most straightforward account of the co-occurrence of negative inversion and *not* is to propose that *not* in inverted clauses is the adverbial NPI minimiser *not* representing stage two of the Jespersen Cycle. Under this hypothesis, the loss of negative inversion in clauses with *not* follows from competition between the minimiser *not* at stage two, and the negative marker *not* at stage three. Some instances of *not* in inverted clauses occur with scalar predicates such as *kepen* 'care for/desire' (77), paralleling the distribution of OE *nawiht*.

[13] The distributions of *na* in inverted and uninverted clauses are significantly different in the periods 850–950 ($\chi^2$(1df)=13.12, p<.001) and 950–1050 ($\chi^2$(1df)=10.72, p<.001). However, this difference becomes non-significant in the eleventh century as *na* becomes equally frequent in both contexts.

[14] Although this analysis predicts that the frequency of ME *ne...not* will be higher than the frequency of OE *ne...na* in uninverted clauses, because ME *not* is reanalysed as a negative marker while OE *na* is not; it does not follow that the frequency of ME *not* is higher than the frequency of OE *na* in inverted clauses. In Section 4.4, I argue this difference derives from a greater overlap between the contexts for negative inversion and the adverb *not* in ME than for negative inversion and the adverb *na* in OE.

[15] The syntactic positions available to *not* also differ in inverted and uninverted clauses. In fact, the three types of *not* identified in this chapter – NPI minimiser, focus adverb and negative marker – are all syntactically distinct in terms of the positions available to them in the clause structure, as I describe in Section 5.3.

(77) Meiden qð he Margarete, ȝet ich bidde & bodie, þat tu
Maiden said he Margaret, yet I bid and command that you
wurche mi wile & wurðgi min maumez, & þe tide & time
do my will and honour my idols and the season and time
schal beon iblescet þt tu ibore were. Nai qð ha ne kepe ich
shall be blessed that you born were. No said she NEG care I
nawt þt me blesci me swa ac hit were þi ȝein & þi god
not that me bless me so but it were your gain and your good
baðe þt tu þe geast unblescet efter blescunge ga
both that you that go unblessed after blessing go
'Maiden, said he, Margaret, still I bid and comand that you do my will and honour my idols and the time you were born will be blessed. No said she I do not care to be so blessed but but it would be both your gain and to your good that you who are unblessed go looking for blessing...'
(CMMARGA,84.463)

However, not all occurrences of *not* in inverted clauses can be explained this way. In the examples given in (78), we cannot interpret *not* as a minimiser with the predicates *bidden* 'ask' or *tellen* 'tell'. Neither of these denote activities or states that are quantifiable by degree or extent, unlike *kepen* 'care' in (77).

(78) a. ne bidde ich nawt drihtin þis for deaðes drednesse. Ah
NEG ask I not Lord this for death's fear, but
false swa hare lahe & festne i þine icorene treowe
prove-false so their law and affirm in your chosen true
bileaue
belief
'I do not ask this, Lord, out of fear of death, but so as to prove their law false and affirm true faith in your chosen ones'
(CMJULIA,122.465)

b. Ne mai ich noht alle þing tellen. ne ich ne can here alre
NEG can I not all things tell nor I NEG can here all
name nemnen. Ac alle woreld þing ben fleted alse water
names name but all worldly things are fleeting as warter
erninde
running
'I cannot tell all the things, nor here mention all their names but all worldly things are fleeting like running water'
(CMTRINIT,177.2407)

While some instances of adverb *not* in ME inverted clauses, such as (77), can be analysed as adverbial minimisers, not all can. In ME inverted clauses the distribution of *not* is not restricted to scalar predicates where it functions as a minimiser. So what are the constraints on its use? Van Kemenade (2011) argues that OE *na* in inverted clauses functions as an adverbial particle, marking contrastive focus. However, she does not extend this analysis to early ME *not*. Instead, she regards ME *not* as a negative marker in both inverted and uninverted clauses. This treatment is at odds with the hypothesis I propose here in which the negative marker *not* and negative inversion are mutually exclusive.

Early ME contexts of negative inversion typically involve contrastive focus, for example (79). Both *na* and *not* appear in these contexts in twelfth- and thirteenth-century texts. In (79), the focus adverb *na* survives and is used much like it is in OE.

(79) Nis na þe halia gast wuniende on his icunde. swa se he iseȝen
NEG-is not the holy ghost existing in his nature as that he seen
wes; forðon þet he is uniseȝenlic. Ac for þere itacnunge swa we
was because that he is invisible but for the sign as we
er seiden. þet he wes iseȝen on culfre and on fure.
before mentioned that we was seen in dove and in fire
'The Holy Ghost is not in his nature existing as he was seen, for he is invisible; but he was manifested in the form of a dove and of fire, for the sign (reason) we have previously mentioned' (CMLAMBX1, 97.86)

In (80), *not* is used in the same way as earlier *na* to mark polarity focus in contrastive constructions where there is a polarity contrast between two conjuncts conjoined by *ac* 'but'.

(80) a. Ne cam ic noht te ȝiuen ȝew for-bisne of mine aȝene wille to
NEG came I not to give you prophecy of my own will to
donne, ac I cam for to donne mines fader wille.
do, but I cam for to do my Father's will.
'I did not come to give you a prophecy of my own will, but to do my Father's will' (CMVICES1,15.161)

b. oðer hwile þe lust is hat towart an sunne. ne macht þu naut
again while the lust is hot towards a sin NEG can you not
þe hwiles deme wel hwat hit is ne hwet þer wile cumen ut
the time judge well what it is nor what there will come out
of. ac let lust ofer gan & eft hit wile þe liken
of but let lust over go and again it will you be glad
'again while desire is hot for any sin you cannot at the time judge well what it is or what will come of it, but let desire pass over and you will be glad.' (CMANCRIW,II.96.1143)

c. ne bidde ich nawt drihtin þis for deaðes drednesse. Ah
NEG ask I not Lord this for death's fear, but
false swa hare lahe & festne i þine icorene treowe
prove-false so their law and affirm in your chosen true
bileaue
belief

'I do not ask this, Lord, out of fear of death, but so as to prove their law false and affirm true faith in your chosen ones'
(CMJULIA,122.465)

d. Cost hehte mi feader ant habbe ihauet hiderto swiðe hehe
Cost was-called my father and have had hitherto very great
meistres Ah forþi þt te lare þt heo me learden limpeð
teachers but beause that the teaching that they me taught leads
to idel ȝelp ant falleð to biȝete to wurdschipe of þe worlde
to idle boasting and falls to beget to worship of the world
ne ne helpeð nawiht eche lif to haben; ne ȝelpe ich nawiht
nor NEG helps not eternal life to have; NEG boast I not
þerof ah sone se ic seh þe leome of þe soðe lare þe
thereof but soon that I see the light of the true teaching that
leadeð to þt eche life, ic leafde all þt oðer…
leads to the eternal life, I abandoned all the other…
'My father was called Cost and until now I have had very great teachers. But because the learning they taught me encourages idle boasting, is aimed at profit and worldly honour, and in no way helps one attain eternal life, in no way do I boast of it. But as soon as I saw the light of the true teaching that leads to eternal life, I completely abandoned the other…' (CMKATHE,26.110–112)

e. for nabbe ich nawt teos niht i worldliche wecchen ah
for NEG-have I not this night in worldly watch but
habbe in heouenliche…
have in heavenly…
'For this night I have not been kept awake in any worldly watch but in a heavenly…'
(CMKATHE,41.351)

f. Meiden qð he Margarete, ȝet ich bidde & bodie, þat tu
Maiden said he Margaret, yet I bid and command that you
wurche mi wile & wurðgi min maumez, & þe tide &
do my will and honour my idols and the season and
time schal beon iblescet þt tu ibore were. Nai qð ha ne
time shall be blessed that you born were. No said she NEG
kepe ich nawt þt me blesci me swa ac hit were þi ȝein &
care I not that me bless me so but it were your gain and
þi god baðe þt tu þe geast unblescet efter blescunge ga
your good both that you that go unblessed after blessing go
'Maiden, said he, Margaret, still I bid and comand that you do my will and honour my idols and the time you were born will be blessed. No said she I do not care to be so blessed but it would be both your gain and to your good that you who are unblessed go looking for blessing…' (CMMARGA,84.463)

g. Ic nam noht giet sad of mine sinnes and forþi ne mai
I NEG-am not yet sated of my sins and therefore NEG can
ich hie naht forlete ac oðer ic mei ben sed þeroff and
I them not renounce but another I may be sated thereof and
þanne ic wille hem forleten...
then I will them renounce
'I am not yet sated of my sins and therefore I cannot renounce them, but another time I may be sated of them and then I will renounce them' (CMTRINIT,75.1027–28)

We also find *not* used to mark contrast in clauses where a clause-initial element other than *ne* is focused. In (81a), clause-initial *hem* 'them' is focused, and focus marked by V to C movement. The focused object *hem* 'them' in the negative clause is contrasted with *þo* 'those' in the affirmative second conjunct, and *not* marks the contrast between the negative and affirmative conjuncts. Similarly in (81b), clause initial *of hem* 'of them' is focused, the focus marked by V to C movement, and *of hem* contrasted with *of þe gostliche*. Again the contrast between the two conjuncts is one of polarity, and is marked by *not*.

(81) a. hem ne munegede he naht to þisse fihte. Ac þo he
them ne exhorted he not to this fight but those he
munegede þe haddn heore sinnes forleten and bette, oðre
admonished that had their sins forsaken and repented or
bigunnen to beten
begun to repent
'It was not them he exhorted to this conflict, but he admonished those that had forsaken their sins, repented or begun to repent' (CMTRINIT,187.2588)

b. þe lichamliche wedes ben manie kinnes ac of hem ne
the bodily garments are many kinds but of them NEG
speke ich noht ac do of þe gostliche
speak I not but do of the spiritual
'the bodily garments may be of many kinds, but I do not speak of them but rather of the spiritual'
(CMTRINIT,95.1272)

Even in contexts where contrast is not marked syntactically by *ac* 'but', *not* may be used in coordinated constructions to mark a contrast between the conjuncts, as in (82).

(82) a. Nart þu naut þe forme ne þe leste
NEG-are you not the first nor the last
'You are not the first, nor the last' (CMANCRIW,II.69.766)

b. ȝef ha beoð nech. ne for hoȝe ich ham nocht. & þach
If they are near NEG care I them not and though
ha beon feor naut i ne reche
they be far not I NEG care
‘ [Concerning smells...] I am indifferent, whether they are nearby or far away’ (CMANCRIW,II.83.1008)

In some other instances where negative inversion and *not* co-occur the function of *not* is less obvious. *not* may mark contrast outside coordinated constructions. In (83), the negative proposition clearly contrasts with and denies its immediate antecedent. Similarly, in (84), the negative clause acts as an explicit denial of an antecedent clause.

(83) & is of dred leost god habbe hire al for warpan. naut nis hit
and is afraid lest God has her all away cast. Not NEG-is it
swa.
so.
‘and [she] is afraid lest God has cast her away completely. No, it is not so.’
(CMANCRIW,II.161.2219)

(84) Ah ȝette me anhwett þt tu ne meahte nawt wearne wið rihte
But grant me which that you NEG might not refuse with right
ȝef me is ileuet þurh mi leoue lauert forte leggen ham adun
if me is given through my beloved Lord forth lay them down
þt þu þi mis-bileaue lette þanne lanhure & lihte to ure.
that you your wrong belief leave then indeed and come to ours.
Nai qð he hetterliche as him þt hoker þuþte ne lið hit
No said he fiercely as him that insulted considered NEG lies it
nawt to þe to legge lahe up-o me…
not to you to lay law upon me…
‘ “But grant me one thing, which you cannot justly refuse: that if it is given me, though my beloved Lord, to lay them low, then at least you will leave your wrong belief and adopt ours.” “No!” he said fiercely, like one who considered himself insulted. “It is not for you to lay down the law for me.…” ’
(CMKATHE,31.191–2)

In (85) there is a polarity contrast between the two subordinate clauses. The proposition that is not understood and not observed during the fast *þat alswa michel senne hit is to breken fasten mid drinke after none wið-uten michele nede* ‘that it is as great a sin to break the fast with drink after noon, unless through great need’ is clearly contrasted with the clause that we can see from the preceding context *is* understood and observed *alswa hit is to-foren non of aten wið-uten alswa michele niede* ‘as it is before noon to eat without also great need’.

(85) Þanne hie alles fasten sculen, ðane fasteð hie þat none
Then they all fast ought, then fast they until noon
uneaðe ðanne after non drinkeð all daiȝ and sume ȝiet
with difficulty then after noon drink all day and some also
benihte. Ne understandeð hie naht þat alswa michel senne hit is to
at night. NEG understand they not that truly much sin it is to
breken fasten mid drinke after none wið-uten michele nede, alswa
break fast with drink after noon without much need, as
hit is to-foren non of aten wið-uten alswa michele niede.
it is before noon to eat without also much need.
'Then they ought all fast, then they fast until noon with difficulty and after noon drink all day and some also at night. They do not understand that it is as great a sin to break the fast with drink after noon, unless through great need as it is before noon to eat without also great need'
(CMVICES1,137.1701)

In (86) again the contrast is implicit. The most appropriate reading in the context is that 'you are not alone in doing so – so do many others, and that they do much worse' rather than 'you do not do so…'.

(86) ne gost þu naut þe ane. monie deð muche wurse
NEG go you not you one. many do much worse
'You are not alone in doing so. Many do much worse'
(CMANCRIW,II.69.769)

The exceptions to these contrastive focus-based uses of *not* in inverted clauses – clauses in which inversion and *not* co-occur outside contexts of contrast or coordination (whether grammatically marked or not) – are clauses in which the verb has a subjunctive or imperative form, or an irrealis or imperative reading, as in (87). Given that inversion in subjunctives and imperatives occurs irrespective of their polarity, inversion in these clauses is most likely motivated by subjunctive or imperative mood rather than negative polarity. Inversion and the negative marker *not* are not incompatible in imperative or subjunctive clauses, and therefore *not* is potentially a negative marker in these clauses.

(87) ȝe leafdis & ȝe meidnes ȝef ȝe weren wise nalde ȝe nawt
You ladies and you maidens if you were wise NEG-would you not
bringe me forð towart blisse wið se bale bere.
bring me forth towards heaven with such tearful outcry
'You ladies and you maidens, if you were wise would not conduct me to heaven with such a tearful outcry'
(CMKATHE,51.500)

The examples presented in this section show that, in contexts of negative inversion, the adverb *not* marks contrastive focus. In Section 5.3.3, I also show that the syntactic positions available to *not* in inverted clauses differ from the position of the negative marker in uninverted clauses in a way that is consistent with analysing *not* as a focus adverb in inverted clauses.

### *4.4.2* Not *in Uninverted Clauses*

Labrum (1982) argues that *not* first emerges in contexts of contrastive focus, where focus is marked explicitly (for example in the NOT X BUT Y construction), generalising to non-contrastive contexts via contexts in which contrast is implicit. What evidence is there that *not* starts out as a contrastive focus adverb in uninverted clauses and is reanalysed as a negative marker? Is this reanalysis ongoing in EME, and is it responsible for the loss of negative inversion?

In EME uninverted clauses, *not* is not restricted to contrastive environments like it is in inverted clauses. Table 4.9 reports the frequency of *not* in uninverted contrastive and uninverted non-contrastive contexts in Early Middle English. In Table 4.9, contrastive contexts comprise a negative clause conjoined with an affirmative clause using the conjunction *ac* 'but', as in (88).

(88) And se prior and se munecas of Cantwarabyrig & ealle þa oðre
And the prior and the monks of Canterbury and all the others
þe ðar wæron munechades menn hit wiðcwæðan fulle twa dagas
that there were monks' men it resisted fully two days
ac hit naht ne beheld
but it not NEG held
'And the prior and the monks of Canterbury and all the other monks' men who were there resisted it fully two days but it did not hold'
(CMPETERB,43.51)

Table 4.9 shows that the frequencies of *not* in contrastive and non-contrastive contexts converge by the mid fourteenth century. The difference between contrastive and non-contrastive contexts is significant in 1150–1250 ($\chi^2$(1df)=43.6, p≤.001.), but not in 1250–1350. Is this convergence due to reanalysis of the contrastive focus adverb *not* as a negative marker?

Table 4.9. *The distribution of early Middle English not in contrastive and non-contrastive environments. Uninverted clauses only.*

| | 1150–1250 | | | 1250–1350 | | |
|---|---|---|---|---|---|---|
| Context | *not* | Total | % *not* | *not* | Total | % *not* |
| Contrastive | 79 | 127 | 62.2% | 61 | 77 | 79.2% |
| Non-contrastive | 153 | 501 | 30.5% | 416 | 550 | 75.6% |

Table 4.10. *Results of mixed-effects regression for predicate type in Early Middle English.*

| Period | Contrastives | p |
|---|---|---|
| 1150–1250 | 0.874 | <.001 |
| 1250–1350 | 0.782 | .02 |

A mixed-effects regression analysis of the data in Table 4.9 addresses the question of reanalysis. If reanalysis of a focus marker as a sentential negator is ongoing in EME, the likelihood of finding *not* in contrastive contexts will lessen over time as the overall frequency of *not* increases. Table 4.10 compares the results of two regression analyses, one for the period 1150–1250, the other for the period 1250–1350. Corpus text is a random effect and contrastive/non-contrastive a predictor variable, with the reference level non-contrastive.

Table 4.10 shows that contrastive contexts consistently favour *not* irrespective of the changing overall frequency of *not* across the two periods. Therefore, the spread of *not* in both contexts proceeds at an approximately constant rate. The small difference between these two effect sizes is likely to be due to chance on datasets of this size, c.600 tokens in each period. These regression results provide no evidence for ongoing reanalysis of contrastive adverb *not* as negative marker *not* during EME. Instead, the regression analyses point to an scenario in which *not* in uninverted clauses is a sentential negative marker from at least the twelfth century onwards. The consistency of the contrastiveness constraint suggests simply that, as it spreads, the sentential negative marker *not* is favoured in contrastives over non-contrastives in EME. Two conclusions follow: (a) that *not* in inverted clauses is functionally distinct (a focus adverb) from *not* in uninverted clauses (a negative marker) in EME; and (b) that the loss of negative inversion in EME results from the spread of a pre-existing negative marker *not* and the consequent loss of $ne_1$, rather than from the reanalysis of *not* from adverb to negative marker during the twelfth and thirteenth centuries.

Maybe *not* starts out marking contrastive focus in Old English and is reanalysed as a negative marker prior to EME. In OE uninverted clauses, 35.5% (n=11/31) of instances of OE *not* appear in contrastive contexts[16], whereas in inverted clauses 63.6% (n=7/11) of *not* appear in contrastives like (89).[17]

16 This is comparable with the period 1150–1250 in which 34.1% (n=79/232) of *not* appear in contrastive environments.

17 *na* appears much more frequently in negative inverted clauses than *nawiht*. *na* appears in 26.5% (n=112/423) of clauses with negative inversion. *nawiht* appears in only 2.6% (n=11/423) of clauses with negative inversion.

(89) a. Ða cwæð ic: ne lyst me nawiht ðara metta þe ic
Then said I: NEG desire me not the meat that I
forhatan habbe. Ac me lyst ðara þe ic getiohhod to
renounced have, But me desire those that I determine to
ætanne ðonne ic hi geseon
eat when I them see
'Then I said "I do not desire those meats that I have renounced, but I desire those that I think right to eat when I see them".'
(cosolilo,Solil_1:38.1.496)

b. Ne derode Iobe naht þæs deofles costnung. ac fremode.
NEG hurt Job not the devil's temptation. but profited.
for ðan ðe he wæs fulfremedre on geðincðum and Gode near
because he was perfect in honours and God near
æfter þæs sceoccan ehtnysse
after the devil's persecution
'The devil's temptation did not hurt Job but profited him because he was more perfect in honours and nearer to God after the devil's persecution.'
(cocathom2,ÆCHom_ II,_ 35:262.61.5860)

c. Ða cwæð ic: ne gebelge ic me nawiht wið þe, ac fagnige
Then said I: NEG be-angry I me not with you but rejoice
þæs þu cwyst
that-GEN you say
'Then I said: I am not angry with you but rejoice in what you say'
(cosolilo,Solil_1:36.1.469)

It is clear that the OE adjunct *nawiht* is not restricted to contrastive or focal environments in OE in either uninverted or inverted clauses. The frequent occurrence of OE *not* in non-contrastive uninverted clauses, like (90), does not support the hypothesis proposed by Labrum (1982) that *not* originates in contrastive contexts during the OE period.

(90) He cwæð þa mid wope, me þincð þæt ðu plegast and þu mine
He said then with weeping, me think that you play and you my
yrmðe naht ne gefredst.
misery not NEG feel
'He said then with tears, I think that you play with me and you do not feel my misery'
(coaelhom,ÆHom_ 27:89.3980)

Instead, *not* originates as an adverbial NPI minimiser in Old English. In EME, it becomes a focus adverb in inverted clauses replacing OE *na*, and a negative marker elsewhere.

In inverted clauses *not* is syntactically as well as functionally distinct from *not* in uninverted clauses. van Kemenade (2011) and Haeberli (2011) agree that OE *na* and ME *not* can occupy a high syntactic position (a position higher than TP) in inverted clauses, but only a low position (a position between TP and VP) in uninverted clauses, as shown in (91).[18]

(91)

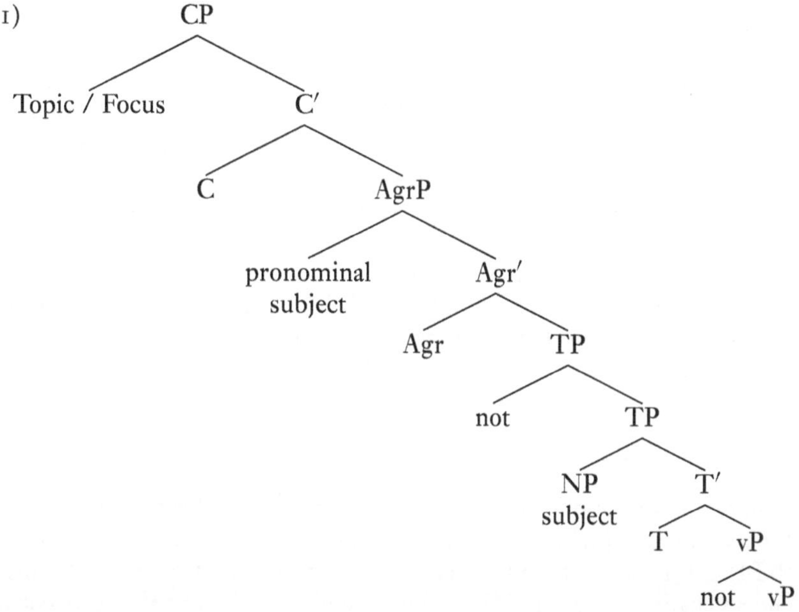


Evidence for the high position comes from the relative positions of *not* and nominal subjects in inverted clauses. Both van Kemenade (2001) and Haeberli (2002a) argue that nominal subjects ordinarily appear in spec,TP[19], and that *not* can precede the nominal subject, indicating that it occupies a position higher than TP. Haeberli and Ingham (2007) discuss example (92) as an example of 'high' *not*.

(92) þenne cumeð þe werse to þe manes heorte and wið his þonc
then comes the devil to the man's heart and with his thoughts
sunderene halt þus queðinde wi sholdest þu þis finden þe
converse hold thus saying why should you this find you
noht ne fost þereof ac he fohð al þat þere comeð he finde
nothing NEG receive thereof but he takes all that there comes, he find
þis nu. Swo ne answerede noht moyses ure drihten þo he bad
this now. So NEG answered not Moses our Lord when he bade
him minister maken
him minister make
'then comes the devil to man's heart and holds private conversation with his thoughts, thus saying: Why should you find this, of which you

[18] For now I assume that *not* is either TP or vP-adjoined, ignoring the question of NegP. This is simply for ease of exposition here. I return to the question of NegP and refine the analysis in Section 5.3.

[19] Except with unaccusative verbs, where subjects may remain in VP.

receive nothing; but he takes all that comes there, let him provide this now. Moses did not thus answer our Lord when he bade him to make a minister' (CMTRINIT,215.3009, translation from Morris (1873)).

Here the first constituent 'So/Thus' is focused ('It was not thus/in that way that Moses answered our Lord...'). It is anaphoric – the preceding clause is its antecedent. The implication that Moses *did* answer, but in a different way to the manner stated in the preceding clause, instead of Moses offering no answer at all to God. Although contrastiveness is not syntactically marked by the conjunction 'but', the negative clause is contrastively focused against the antecedent clause.

The occurrence of high *not* in inverted clauses indicates that it is not a sentential negative marker but a focus adverb. Therefore, the high position for *not* is not the specifier of a functional projection NegP (contra both van Kemenade (2011) and Haeberli (2011)).

## 4.5 Summary

Different distributions of *not* in OE and Early ME indicate there are three functionally and grammatically distinct forms of *not*: an adverbial minimiser (NPI) in OE, which then becomes a focus adverb in EME inverted clauses, and a negative marker in EME uninverted clauses. *Ne*-initial clauses constitute a distinct environment in which the negative marker *not* cannot occur, because clause-initial *ne* is itself a negative marker.

The data presented in Section 4.2 demonstrate that *not* has its origins in OE as an adverbial NPI minimiser, similar in distribution to PDE *one bit*, occurring with scalar predicates. By Middle English, *not* has been generalised to non-scalar predicates, pointing to a date for the reanalysis of the adverbial minimiser *nawiht* as a negative marker prior to the Middle English period, possibly in the tenth or eleventh centuries. Two factors contribute to the reanalysis of *not* – the morphosyntactic weakening of *ne* (which is not confined to contexts with *not*, see Chapter 8), and the structural ambiguity of bipartite *ne...not*. The reanalysis of *not* results in competition between stage two $ne_1$*...not*, in which *not* is an adverb and $ne_1$ is the negative marker; and stage three $ne_2$*...not* in which *not* is a negative marker and $ne_2$ is a concordant negative item. Only stage two $ne_1$*...not* is compatible with negative inversion. The loss of negative inversion is therefore a consequence of competition between the negative markers $ne_1$ and *not*.

Having established that the early English Jespersen Cycle involves two distinct forms of *ne* and three distinct forms of *not*, Chapter 5 proposes a syntactic account of these different forms and their interactions within the Jespersen Cycle within a formal Minimalist framework. I argue that the distributional differences between these elements arise through minimal differences in their morphosyntactic features, and that competing forms are featurally equivalent.

# 5 The Syntax of the Early English Jespersen Cycle

## *A Morphosyntactic Feature-based Account*

It is clear from the preceding chapters that formal accounts of the Jespersen Cycle which assume a single form of *ne* such as Roberts and Roussou (2003), Zeijlstra (2004) do not accommodate the changes in the distribution of *ne* described in Chapters 2 and 3 and therefore fail to model the distribution of *ne* adequately across the Jespersen Cycle. Instead, we need a means to distinguish two forms of *ne*. One way to do this would be to invoke a semantic distinction between a negative marker with a semantic negative feature and a negative polarity item (NPI) without. However, it raises the issue of what properties the NPI *ne* has within a taxonomy of negative polarity items such as that proposed by Hoeksma (2012), and how it is licensed. Breitbarth (2009) proposes to treat ME *ne* as a non-veridical polarity item. Zeijlstra (2010) proposes that French *ne* is also a non-veridical NPI. However, this kind of analysis overgeneralises the distribution of ME *ne*. Instead, Middle English *ne* has the properties of an anti-veridical NPI, in that it must be licensed by negation. Furthermore, the licensing conditions on ME *ne* are subject to strict syntactic locality constraints.

Consequently, while it is plausible that an account could be sketched out in purely semantic terms to account for the changing distribution of *ne*, I instead model the changing distribution of *ne* within a Minimalist syntactic framework based on morphosyntactic features and feature agreement.

## 5.1 Syntactic Framework and Assumptions

The syntactic framework on which the analysis is based is a modified version of the Minimalist framework set out in Chomsky (1999, 2000), incorporating later proposals from Hiraiwa (2001), Pesetsky and Torrego (2007), Wurmbrand (2012) and Zeijlstra (2012). The approach is a modular one, in which the syntactic derivation assembles lexical items into larger structures which are interpreted or spelled out at two interfaces - PF (phonological form), where the morphological and phonological form of the syntactic unit is spelled out, and LF (logical form) where the syntactic unit is interpreted semantically. Semantic compositionality is fundamental, with the meaning of the structure read off its syntactic form. The key principles of this framework that will become relevant here are:

**The syntactic derivation:** Syntax (the derivation) mediates between the lexicon (or a subset of the lexicon, the numeration – lexical items selected for a particular clause) and two interfaces: the sensory-motor interface at which the clause is realised phonologically and morphologically (also known as Phonetic Form or PF); and the conceptual-intensional interface at which the clause is interpreted semantically (also known as Logical Form or LF).

**Morphosyntactic features:** Morphosyntactic features associated with each lexical item (in the lexicon) determine the syntactic distribution and semantic interpretation of that lexical item. Lexical items are form-meaning pairings that are interpreted at the two interfaces PF and LF. Lexical items are feature-bundles. Thus parametric variation reduces to differences in the features that are associated with particular lexical items (differences in the feature specification of those items).[1] Regular patterns of morphosyntactic change suggest that there are particular pathways of change in the feature specifications of lexical items.

**Syntactic dependencies:** Lexical items are licensed by, or enter into syntactic configurations with, other lexical items only when their morphosyntactic features require it. Hence differences in the syntactic configurations and dependencies lexical items enter into result from differences in the morphosyntactic feature specifications associated with them in the lexicon.

**Locality:** the syntactic structure is built up in phases in the sense of Chomsky (1999), with cyclic spellout at the interfaces. The derivation builds syntactic structure in chunks which are then interpreted at the interfaces, meaning that these already spelled-out chunks are no longer visible to further syntactic operations. This results in locality constraints.

Changes in the feature specifications of lexical items result in changes in the syntactic distributions of those items. Most Minimalist analyses since Chomsky (1995) have distinguished two types of morphosyntactic features, one type that contributes meaning to the C-I interface, the other which exists purely to derive syntactic relationships between lexical items within the clause.

Another important point for our purposes is that the semantics of LF are compositional. For a clause to be interpreted with clausal scope negation,

[1] As Adger and Svenonius (2010, 3) claim, within a feature-driven Minimalist framework '...entertaining alternative hypotheses about feature structures is tantamount to entertaining alternative theories.'

it can have at most one semantically interpretable negative feature in its derivation, in a position of sentential scope. Thus when a clause contains more than one negative item, for example in the bipartite *ne...not* stages of the Jespersen Cycle, one or other of *ne* or *not* can have a semantically interpretable negative feature associated with it, but not both.[2] This raises the question of what the feature-specifications of the other (concordant) negative items are, and how the negative concord reading arises.

## 5.2 Morphosyntactic Features and the Jespersen Cycle

Structuring the Jespersen Cycle in terms of variation and change in the morphosyntactic feature specifications of lexical items is not a new idea (cf. Wallage (2005), Wallage (2008), van Gelderen (2008), Breitbarth (2009), van Gelderen (2011: 299), Willis (2011a)). Within such an account, the difference between *ne*$_1$ and *ne*$_2$ is minimal and arises out of a difference in feature specification which can be parameterised simply as choice between two featurally distinct but homophonous items within the lexicon. The analysis thereby follows the Borer conjecture (Borer, 1984) that parametric variation reduces to lexical variation.

Adger and Svenonius (2010) describe different approaches to features within various Minimalist approaches. There are (at least) two different ways to conceive of the second order distinctions between types of features within different implementations of the framework.

**Chomsky (1995):** In Chomsky (1995), features are marked as semantically interpretable [*i*F] or semantically uninterpretable [*u*F]. Features interpretable at LF have an an effect at the C-I interface. Features uninterpretable at LF can be thought of as features which have an overt morphological or phonological reflex at PF but which are deleted by syntactic operations prior to semantic interpretation at LF. The sole function of uninterpretable features is to initiate dependencies between lexical items, in order that the syntax can accommodate agreement relations between lexical items.

**Chomsky (1999, 2000):** In Chomsky (2000) and subsequent versions of the theory, features take the form of attribute:value pairings. A distinction is made between valued features which enter the derivation with both an attribute and a value, and unvalued features, which enter the derivation with just an attribute, and receive a value during the syntactic derivation.

[2] There is no negative absorption in this model of the type proposed by Haegeman and Zanuttini (1996). However, there is also the possibility that neither NI hosts the negative feature which marks the clausal scope negation at LF, but that this feature is hosted on some other element, which may be lexically or morphologically covert.

Subject-verb agreement illustrates how this works – both the subject and the verb have $\phi$-features (person and number), but these are only semantically interpretable or valued on nominal elements (the subject). They are unvalued on the verb. The role of this unvalued feature on the verb is to ensure that the verb enters into a syntactic relationship (or dependency) with a subject. The person and number specification of that subject will then determine what person and number morphology the verb has. Valued features do not need to enter into dependencies with any other features in order to be interpreted at LF, whereas unvalued features (features that LF cannot interpret because they are in some sense incomplete) need to be eliminated from the derivation before LF is reached. They are deleted after they enter into a dependency with a valued feature of the same type. Feature checking and deletion can take place via Merge – insertion of a constituent at the root of the derivation, thus extending the syntactic structure; or via Agree. Agree operates within the existing syntactic structure, so does not extend the syntactic structure. The precise nature of the Agree relation is a matter of debate that has empirical consequences (for discussion see Section 5.4).

As an uninterpretable or unvalued feature enters into a syntactic dependency with a matching interpretable or valued feature, whether or not a lexical item has interpretable/valued or uninterpretable/unvalued features parameterises aspects of its syntactic distribution. Up to this point, I have assumed that feature-interpretability and feature-valuation are equivalent. This is not necessarily the case. The interpretable – uninterpretable distinction suits negative features well in the sense that an interpretable negative feature corresponds to semantic negation ¬, while an uninterpretable negative feature is associated with a word that is negative in form but does not introduce a negative operator ¬ at LF.

However, Breitbarth (2009) argues that, within a valued/unvalued feature model, making a distinction between a valued and unvalued negative feature is conceptually odd – negative seems conceptually more like a value than an attribute. This leaves open the issue of what the attribute is that can be valued negative. Negation is one of a class of affective or non-veridical operators, such as the operators that license the *any-* series of NPIs in the PDE sentences in (93). (94) shows that Present-day French *ne* appear in similar set of non-veridical contexts.

(93) a. I didn't see anything (negation)
b. If there is anything I can do, please let me know (conditional)
c. John is taller than anyone else I know (comparative of inequality)
d. He left before I said anything (*before* clause)

(94) a. Jean ne mange pas
John ne eats not
'John does not eat' (negation)

b. Je viendrai à moins que Jean (ne) soit là
I will-come to less that Jean (ne) is.subj there
'I will come unless Jean is there' (conditional)
(Zeijlstra, 2010, ex.8f)

c. Jean est plus malin que Pierre (ne) l' est
Jean is more smart than Pierre ne it is
'Jean is smarter than Pierre is' (comparative of inequality)
(Zeijlstra, 2010, ex.8b)

d. Il est parti avant que nous (n') ayons mange
He is left before that we ne have eaten
'He left before we ate' (*before* clause)
(Zeijlstra, 2010, ex.8g)

Breitbarth (2009) argues that stage three negative heads, like French *ne*, have an unvalued affective feature [aff: ], valued by Agree with an affective operator, which has a valued Aff-feature (for instance, negation is valued negative [Aff:neg]). While this may capture the distribution of Present-day French *ne*, licensed across the range of affective contexts in (94)[3], Breitbarth's account overgeneralises the distribution of ME *ne*.[4] The problem is that an unvalued [Aff:__] feature can be valued by a number of affective operators, not just negation. This predicts ME $ne_2$ will appear in all the affective contexts in (93) and (94) – that it has the distribution of a non-veridical NPI. However, we saw in Chapter 3 that ME $ne_2$ is an anti-veridical negative item licensed only in concord with a negative operator. Even some concordant negative items that do not contribute negative force at LF are still identified by speakers as negative, because they only appear in dependencies with other negative items. We need an account that will restrict ME $ne_2$ to negative environments.

Pesetsky and Torrego (2007) argue that the two dimensions of feature interpretability and feature valuation are not exclusive. This provides the three-way contrast in (95) between OE $ne_1$, ME $ne_2$ and modern French *ne*.[5]

(95) Old English $ne_1$: [iAff:neg]
affective operator, valued negative in the lexicon, LF-interpretable – negative marker.
Middle English $ne_2$: [uAff:neg]
inherently or lexically valued negative, but semantically uninterpretable – concordant negative item/anti-veridical NPI. Agrees with [iAff:neg] only.

[3] Negative concord contexts are somewhat more problematic, suggesting perhaps that the licensing mechanism for *ne* is semantic rather than syntactic (see Zeijlstra (2010)).
[4] Similarly, Biberauer and Roberts (2011) argue that an operator feature [Op:] may have the value negative. This potentially overgeneralises the distribution of ME *ne* in the same way.
[5] It is not clear what the fourth option [iAff:__] would correspond to in this particular instance since an operator cannot be interpreted at LF if we do not know what kind of operator it is (that is, if it is unvalued). However, Pesetsky and Torrego (2007) present arguments for LF-interpretable unvalued features elsewhere in the grammar.

Present-day French *ne*: [uAff:__ ]
unvalued affective feature, valued by any affective (non-veridical) operator – non-veridical NPI. Agrees with any valued [iAff:] feature, irrespective of its value.

This hybrid approach may be a way to accommodate the different distributions of early English *ne* and French *ne*, perhaps a matter of parametric variation,[6] or even as successive diachronic stages within the Jespersen Cycle.[7] In this account, ME $ne_2$ may represent an intermediate stage in the weakening of *ne* in which *ne* has lost its LF-interpretability, but not its value. On the other hand, it may be better to license French *ne* semantically rather than syntactically: the taxonomy in (95) may not be sufficiently nuanced to accommodate the differences between all the types of polarity items discussed in Hoeksma (2012).

Given the possibilities made available by Pesetsky and Torrego's (2007) interpretation of the Minimalist framework and the distribution of ME *ne* as a concordant negative item (anti-veridical polarity item), it seems most straightforward to hypothesise that the difference between $ne_1$ and $ne_2$ is in the interpretability of a negative feature rather than a difference in its value. The result is that $ne_1$ can mark negation independently at PF and LF, but $ne_2$ marks negation only at PF and so must occur in a syntactic relationship with another (semantically interpretable) negative item, for the derivation to converge at LF. This means that *ne* is only licensed in negative concord.

The result is that the Jespersen Cycle is formalised as (96).

(96) Stage One (OE): $ne_1$ [*i*Neg]
Stage Two (OE): $ne_1$ [*i*Neg] ...$not_1$ [*u*Neg]
Stage Three (ME): $ne_2$ [*u*Neg] ...$not_2$ [*i*Neg]
Stage Four (14th century onwards): $not_2$ [*i*Neg][8]

6 The taxonomy in (95) makes available three different pathways through the Jespersen Cycle: [iAff:neg]>[uAff:neg] (early English), but also [iAff:neg]>[uAff:neg]>[uAff:__] or [iAff:neg]>[uAff:__]. In the former, loss of LF-interpretability and loss of value are independent. In the latter they are simultaneous. The existence of ME *ne* suggests that the loss of interpretability and the loss of value are independent, successive stages, but further research on the diachrony of French – to identify whether there are two forms of *ne* in competition or three – is required to test this.

7 This is a point of difference between languages, for example Present-day French and early English. This difference may arise in two ways. It may reflect a different change in the transition from stage two to stage three of the Jespersen Cycle (in which case we might ask why languages differ in this way), or it may be that languages like French have an additional bipartite *ne...pas* stage in their Jespersen Cycle during which the anti-veridical NPI *ne* is reanalysed as non-veridical. Modelling the constraints on French *ne* over time in the same way we did for English in Chapter 2 might provide evidence to distinguish whether a model with two or three forms of *ne* provides a best fit for the changing distribution of French *ne*.

8 Zeijlstra (2008) makes a distinction between formal (syntactic) and semantic negative features. In a system (such as Zeijlstra's) where negative concord is syntactic agreement, formal (syntactic) negative features are required for a negative item to participate in negative

Descriptive accounts of the Jespersen Cycle, such as Jespersen (1917), make reference to the weakening of *ne*. The proposed account allows each stage in this weakening process to be formalised in terms of morphosyntactic features, first through the loss of feature interpretability, then the loss of feature value: negative operator *ne* [iAff:neg] > anti-veridical NPI *ne* [uAff:neg] > non-veridical NPI *ne* [uAff:__]. Chapters 2–4 argued that the distribution and syntax of ME *ne* differs at stages one and two versus stage three.[9] This follows from competition between two featurally distinct forms of *ne* – [iNeg] at stages one and two versus [uNeg] at stage three. When *ne* co-occurs with *not* it is structurally ambiguous. at stage two, *ne* is the LF-interpretable negative marker just as at stage one of the Jespersen Cycle, and *not* a semantically uninterpretable-neg adverb in concord with *ne*. At stage three, *not* is the LF-interpretable negative marker and *ne* a semantically uninterpretable-neg item in concord with *not*. Chapter 4 described changes in the distribution of *not* with respect to negative inversion during the course of OE and ME. These follow from the distinction between stage two *not* [uNeg] and stage three *not* [iNeg].

This account differs from the feature-based approach proposed by Zeijlstra (2004). Although his approach makes use of the interpretable–uninterpretable feature distinction, he links feature interpretability to syntactic positions such that negative operators in spec,NegP are [iNeg] whilst $Neg^0$ and all other negative items are [uNeg]. However, positing a single form of *ne* throughout the Jespersen Cycle does not allow us to model changes in the distribution of ME *ne* in a way consistent with either its distribution in diachronic data, or with models of morphosyntactic change as grammar competition.

## 5.3 The Positions of Negative Markers in the Clause Structure

Feature-checking under Agree requires items with matching features to enter into particular syntactic configurations. Therefore, ascertaining the positions of *ne* and *not* in the clause at each stage of the Jespersen Cycle is a prerequisite to describing the Agree relation between negative items at each stage of the cycle.

### *5.3.1 The Functional Projection NegP in Feature-driven Analyses*

Within Minimalism, the distinction between categorial features (i.e. a feature of a lexical item which can head a syntactic projection) and non-categorial features (i.e. a feature which is present on a lexical item but does not head its own functional projection) determines how articulated the functional

concord. As I adopt an analysis of negative concord as syntactic agreement, and *not* at stage four can participate in negative concord, I assume that it retains formal [iNeg] features.

[9] Chapter 6 will argue that there is also a functional distinction between these two types of *ne*.

structure of the clause is. Under what conditions do features project, and what does this mean for the representation of negation in clause structure? Many Government-Binding approaches to negation, such as Haegeman (1995) propose that negation heads its own X'-theoretic functional projection. However, the status of NegP is not so clear within Minimalist approaches. For example, Breitbarth (2014) proposes to eliminate NegP entirely.

If we admit variation between *ne* [iNeg] and *ne* [uNeg], *ne* does not always enter into a specifier-head relationship with a negative operator. At stages one and two of the Jespersen Cycle, *ne* is the negative operator. Although it can appear in negative concord with other negative items, it is not required to do so, and can be interpreted at LF as a negative operator independently of any other negative items. At stage two, the concordant negative NPI *not* enters into negative concord with the negative marker *ne*. There is only overt evidence that *ne* enters into a dependency with a negative operator at stage three of the Jespersen Cycle. At this stage, *ne* is a negative agreement marker on the finite verb. The status of *ne* is similar to that of other agreement morphology (such as $\phi$-features which are uninterpretable on the finite verb and its associated functional heads). Given that NegP has the same status as AgrP at stage three of the Jespersen Cycle – its head has only an uninterpretable feature – whether we postulate NegP at stage three depends on whether uninterpretable features project or not.

The representation of agreement morphology depends on the view taken of the morphology-syntax mapping. Chomsky (2000, 138–9) argues that heads consisting entirely of uninterpretable features cannot project:

> In MP [The Minimalist Program] it is speculated that categories lacking interpretable features should be disallowed - specifically Agr, consisting only of uninterpretable $\phi$ features. That conclusion is forced in this version. Suppose is an LI that consists of uninterpretable features only and selects, yielding the syntactic object K=$\alpha$, $\beta$, with label $\alpha$. In the course of a convergent derivation, $\alpha$ will disappear, leaving K and higher projections of without a label. But terms without labels are not well-formed syntactic objects. Accordingly such elements as Agr not only might not exist, but cannot exist, on rather plausible assumptions.

If we substitute *ne* [uNeg] for $\alpha$ and *not* [iNeg] for $\beta$ here, the problem for NegP becomes clearer. Chomsky's proposal constrains the syntactic representation of negation at stage three of the Jespersen Cycle – the [uNeg] feature must be hosted by some other head which also has LF-interpretable features. Checking must be carried out in a multiple specifier configuration. A good candidate to host [uNeg] might be $v^0$. Chomsky (1999) suggests that vP can have more than one specifier, the inner specifier an A-specifier, the outer one an A'-specifier. This formalises the semantic idea that negation scopes over propositions. In the syntax, negation appears at the phase edge. This aligns it with other affective operators such as wh-operators, and suggests a new perspective on weak islands. Weak islands arise at phase edges if the outer specifier of vP is not available for movement or reconstruction.

However, Chomsky's proposal is not the only perspective from which to view the question of NegP. Bobaljik (1995) and Bobaljik and Thrainsson (1998) take each morpheme to correspond to a functional head in the syntax, continuing the Government-Binding approach to functional projections. This is the approach which underpins Haeberli's clause structures for early English. If I admit AgrP on the basis of overt agreement morphology as Haeberli (2002a) does, following Bobaljik and Thrainsson (1998), I am forced to admit NegP on the basis of the negative morpheme *ne*. A similar approach follows from Giorgi and Pianesi's Feature Scattering Principle (97).

(97) Feature Scattering Principle: Every feature can head a projection (Giorgi and Pianesi, 1997, 231)

Nash and Rouveret (1997) argue that uninterpretable features are hosted parasitically on other heads, but that the uninterpretable features may project a proxy projection. Both NegP and AgrP may be analysed as Proxy Projections of $v^0$ and $T^0$ respectively, created for the purpose of checking features only. Under both these approaches the [uNeg] feature on *ne* will project $Neg^0$ solely to provide a specifier position in which the specifier can check and delete [uNeg] on $Neg^0$. These approaches eliminate multiple specifiers, as each feature that is valued heads its own functional projection. Given these different theoretical approaches, the question of NegP becomes an empirical one – whether at each stage of the Jespersen Cycle negative markers occupy positions within the clause that are best derived by hosting them within a functional projection NegP distinct from vP and TP.

### 5.3.2 *Old and Middle English Clause Structure*

Haeberli (2002a) argues for a split-IP structure in Old English. Finite verbs move to $T^0$ in Old English subordinate clauses, and to a higher $Agr^0$ position in main clauses. In this clause structure, there are two subject positions, spec,TP for nominal subjects, and spec,AgrP for pronominal subjects (which represent bundles of $\phi$-features).[10] The resulting clause structure is shown in (98).

[10] However, negative clauses are one of the few clause types (see Pintzuk (1999)) which exhibit subject-verb inversion with both nominal subjects in spec,TP and pronominal subjects in spec,AgrP. This indicates that finite verbs procliticised with *ne* can also appear in $C^0$. For detailed discussion of these clauses, see Chapter 4.

(98)

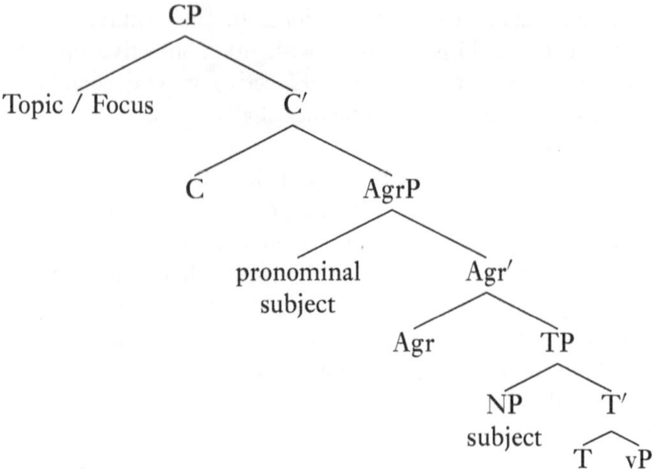


### 5.3.3 *The Position of* not *in Clausal Structure: Upwards Structural Reanalysis at Stage Three of the Jespersen Cycle*

Recall from Chapter 4 that *not* is reanalysed from a complement of $V^0$ to a vP-adjunct at stage two of the Jespersen Cycle. In OE, its distribution is that of an anti-veridical polarity item. It is a minimiser like the PDE NPI *one bit*. An uninterpretable negative feature [uNeg] ensures that it appears in a concord relationship with the negative marker *ne* [iNeg].

The position of OE *not* relative to scrambled objects does not distinguish it from the position of vP-adjoined adverbs. In OE we find examples of full NP objects scrambled to a position structurally higher than *not* – in (99) the NP object appears to the left of *not*, indicating that scrambling has occurred.[11]

(99) a. & ic þine tintregu naht ne gefrede.
and I your torment not NEG feel
'and I do not feel your torment at all'
(cocathom1,ÆCHom_I,_29:423.142.5749)

[11] By the same measure, examples like i. with a full NP object preceding OE *na* suggest that the position of *na* does not differ from the position of vP-adjoined adverbs in uninverted clauses, contrary to claims that it occupies a low spec,NegP position made by van Kemenade (2001) and van Kemenade (2011).

i. Þa ða awyrgdan gastas, þe ðær stodon & se seoca man hefelice aræfnede,
Then the cursed spirits that there stood and the sick man grievously suffering
ne mihton þa oðre men na geseon
NEG could the other men not see
'Then the cursed spirits that stood there and the sick man grievously suffering could not see the other men'
(cogregdC,GDPref_and_4_[C]:40.326.3.4904)

b. He cwæð þa mid wope, me þincð þæt ðu plegast and þu
He said then with weeping, me thinks that you play and you
mine yrmðe naht ne gefredst.
my misery not NEG feel
'He said then, with tears, I think that you play and that you do not feel my misery'
(coaelhom,ÆHom_27:89.3980)

c. Ða cwæð Martianus mid mycclum graman, swa micel is þin
Then said Martin with much rage, so much is your
drycræft þæt ðu þas dyntas naht ne gefretst
sorcery that you the blows not NEG feel
'Then Martin said with fury, "Your sorcery is so great that you do not feel the blows."
(coaelive,ÆLS_[Julian_ and_ Basilissa]:146.1027)

We find the same pattern with manner adverbs like *lichamlice* 'physically' in (100).

(100) Þa undergeat se preost þæt he ne mihte ðone halgan wer
Then understood the priest that he NEG could the holy man
lichamlice acwellan.
physically kill
'Then the priest understood that he could not physically kill the holy man' (cocathom2,ÆCHom_II,_11:96.153.1992)

In contrast with OE *not*, Haeberli and Ingham (2007) propose that Early Middle English *not* occupies the specifier of a NegP that is low in the clause structure, between TP and vP. Their argument is based on an observation that object pronouns may precede both *not* and vP-adjoined adverbs, whilst nominal objects may precede vP-adjoined adverbs but always follow *not*. (101) shows that pronominal objects can scramble out of their base-merged position within VP, moving to a position higher than *not*, but below $T^0$. (102) illustrates that scrambled NPs move to a position higher than manner adverbs within or adjoined to vP[12] but lower than *not*.[13]

(101) þt ich ne seo hire nawt heonne-forð mare
that I NEG see her not henceforth any-more
'that I will not see her any more'
(CMJULIA, 123.489, Haeberli and Ingham (2007, 16, ex.19))

[12] Unlike Haeberli and Ingham (2007), I distinguish vP and VP. I assume, following discussion of manner adverbials in the syntactic literature on PDE, for example by Hale and Keyser (1993) Levin and Hovav (1995), Adger and Tsoulas (2004), that manner adverbials are associated with agentivity and therefore associated with vP rather than VP. Evidence for this comes from the incompatibility of manner adverbials with unaccusatives and passives.

[13] For the period 1150–1250, Haeberli and Ingham (2007) give the following figures from the PPCME2: in main and subordinate clauses with an auxiliary, 0/8 NP objects precede *not*; in clauses with finite lexical verb only, 0/45 NP objects precede *not*. For object pronouns, in auxiliated clauses 4/7 object pronouns precede *not*; in clauses with a finite lexical verb only 38/38 object pronouns precede *not*.

(102) þu qð ha Keiser nauest nawt þis strif rihtwisliche
You said she emperor NEG-have not this contest equitably
idealet
shared
'You, she said, emperor, have not shared this contest equitably.'
(CMKATHE, 30.187. Haeberli and Ingham (2007, 15, ex.17))

Haeberli and Ingham conclude that scrambled pronominal objects move to a higher position than scrambled NP objects, and that *not* occupies a position distinct from – higher than – the position of manner adverbials. A fully articulated clause structure for EME distinguishes the position of *not* from the position of adverbs, and the position of scrambled pronominal objects from the position of scrambled full NP objects, as in (103). Haeberli and Ingham (2007) propose that NP scrambling is NP adjunction to VP – this corresponds to vP-adjunction in the clause structure in (103).[14] In this structure, pronominal objects move across spec,NegP *not* to spec,AgrOP while full NP objects scramble to a vP adjoined position below NegP.

(103)

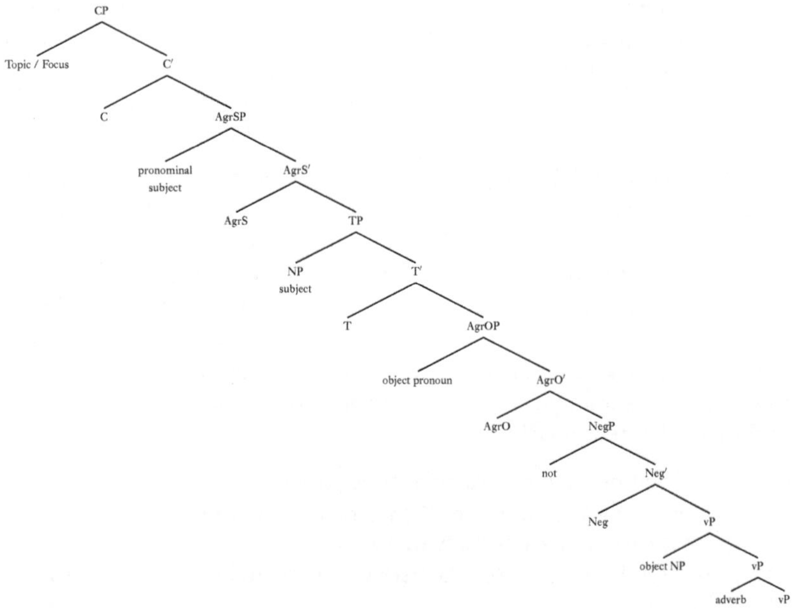


[14] I assume that NegP and AgrOP are proxy projections of vP – that $Neg^0$ and $AgrO^0$ consist entirely of LF-uninterpretable features, and that their specifier positions are projected to facilitate feature checking, as well as to account for the specific word order patterns we find in EME.

Comparison of the OE and EME data point to two conclusions. First, OE *not* appears within or adjoined to vP[15] whereas EME *not* occupies a higher functional position that is distinct from vP adjuncts. Second, it follows that the reanalysis of *not* from OE NPI minimiser to ME negative marker involves the kind of upwards structural reanalysis from lexical to functional positions which has often been proposed in studies of grammaticalisation, for example by Roberts and Roussou (2003).

So far, we have discussed the position of *not* in OE and EME uninverted clauses. The position of *not* in inverted main clauses – that is, clauses with a negated finite verb in $C^0$ – is more variable in PPCME2 data from the period 1150–1250. Section 4.4 argued that when negative inversion and *not* co-occur, *not* is a concordant focus adverbial rather than a negative marker. In clauses with a negated verb in $C^0$, *not* always follows pronominal subjects (n=80/80 inverted clauses with pronominal subjects), but can either precede (n=12/15) or follow full nominal subjects (n=3/15). What constrains this variation? Does the variation result from variability in the position of *not*, or from a fixed position for *not* and variability in the position of nominal subjects?

In many instances, the subjects following *not* are subjects of unaccusative verbs whose subjects may simply remain in VP, such as in (104). The position of these subjects is ambiguous, and not necessarily incompatible with a low position for *not* between TP and VP.

(104) Ne bið naut his lare fremful ne icweme þan ileweden
NEG is not his teaching beneficial nor suitable for the laity
'His teaching is not beneficial nor suitable for the laity'
(CMLAMBX1,109.1006)

In other examples with transitive verbs, such as (105)–(107), the subject (in bold face below) following *not* is within the contrastive focus marked by *not*.

(105) þenne cumeð þe werse to þe manes heorte and wið his þonc
then comes the devil to the man's heart and with his thoughts
sunderene halt þus queðinde wi sholdest þu þis finden þe
converse hold thus saying why should you this find you
noht ne fost þereof ac he fohð al þat þere comeð he
nothing NEG receive thereof but he takes all that there comes, he
finde þis nu. Swo ne answerede noht **moyses** ure drihten þo
find this now. So NEG answered not **Moses** our Lord when
he bad him minister maken
he bade him minister make

[15] By extension of arguments in the literature on PDE syntax, for example Hale and Keyser (1993) Levin and Hovav (1995) Adger and Tsoulas (2004), that manner adverbials are linked to agentivity, and agentivity is encoded on v. The PDE minimiser *one bit* is consistent with this generalisation, in that it does not occur in passive constructions in either the BNC or COCA corpora.

'then comes the devil to man's heart and holds private conversation with his thoughts, thus saying: Why should you find this, of which you receive nothing; but he takes all that comes there, let him provide this now. [but] **Moses** did not thus answer our Lord when he bade him to make a minister' (CMTRINIT,215.3009, translation from Morris (1873))

(106) Nule naut **ure lauerd** þt amon for an þing beo twiȝen
NEG-intends not **our Lord** that a man for one thing be twice
idemed. hit nis naut i godes curt ase i þe schire
judged. It NEG-is not in God's court as in the shire
'**Our Lord** does not intend that a man should be judged twice for the same thing. It is not the same in God's court as in the shire court' (CMANCRIW,II.228.3297)

(107) Ne underuoð nawt qð he þis ilke word **alle**. Hwa-se hit mei
NEG receive not said he this same word **all**. Whoever it may
underneomen, underneome ich reade
receive, receive I counsel
'"**Not all**" he says "receive this same word. Whoever can receive it, receive it, I counsel.' (CMHALI,141.201)

In (105), the subject *Moses* is part of the focalised constituent. Here he is first introduced to the discourse, as new information, and implicitly contrasted with the devil. Similarly, in (106), the subject *our Lord* is new to the discourse. The contrast between the mercy of God and the workings of the temporal judiciary in the next clause indicate that the subject 'our Lord' is within the focus marked by *not* here. In (107), the universal quantifier subject *all* is contrasted with the subject of the following clause, and is within the focus signalled by *not*. However, in (108), the subject *the godfathers* is given information, and as such stands outside the focus marked by *not*.

(108) and **here godfaderes** sullen for hem andswerie bifore þe prest
and **their godfathers** shall for them answer before the priest
ate fanstone... And þat ne mugen þe godfaderes naht don; bute
at font... And that NEG can the godfathers not do; unless
hie here bileue cunnen. and nime to þe children muchele geme.
they their belief know, and give to the children much heed
'**their godfathers** shall answer for them to the priest at the font...and the godfathers cannot do that unless they know their own faith and pay much heed to the children'
(CMTRINIT,17.191)

This distribution of subjects relative to *not* is reminiscent of the distribution of subjects and *na* described by van Kemenade (2011). She claims that subjects new to the discourse follow *na* and subjects given in the preceding discourse (discourse-old) precede *na*.[16] The variable position of nominal subjects relative to *not* indicates that the position of the focus marker *not* is variable, much like other discourse-orientated adverbs, whereas there are fixed positions for both the sentential negator ME *not* (spec,NegP) and the adverbial minimiser OE *not* (vP adjunct).

Finally for this section, we should consider the position of Late Middle English *not* at stage four of the Jespersen Cycle after *ne* has been lost. Does *not* remain a specifier of NegP at this stage or does it become a vP-adjunct? Again, we can examine the position of *not* relative to scrambled objects. Van der Wurff (1997, 1999a) claims that by this time, we can assume that VP is uniformly head-initial. Therefore any OV orders are derived by movement of the object out of VP to a higher position, so both simplex and auxiliated clauses with OV word orders provide evidence for object movement.

In Late Middle English (fourteenth and fifteenth centuries), scrambling of full NP objects is in decline, surviving at any frequency only with negative and quantified objects (for discussion see van der Wurff (1999a)). In LME (1350–1500) clauses with OV word order, both pronominal objects and full NP objects continue to move across adverbs. However, we continue to see the earlier asymmetry between the positions of full NP objects and object pronouns with respect to *not*.[17] The distributions of objects relative to *not* and to adverbs (excluding temporal and locative adverbs) in Late Middle English (1350–1500) auxiliated and simplex clauses are compared in Table 5.1.

These figures may reflect the more general loss of NP scrambling, which van der Wurff (1997, 1999a) argues is largely limited to negative and quantified objects at this time. However, they provide robust evidence of a pronoun–NP asymmetry with respect to object movement across *not*, just as in EME. These data indicate continuity in the position of *not* (spec,NegP) at stages three and four of the Jespersen Cycle. Its position is distinct from vP-adjoined adverbs at both stages. This constitutes indirect evidence for an underlying negative head at stage four. The difference between stages three and four seems to be in the morphological realisation of that negative head – overt *ne* at stage three, null at stage four. There is no change in the position of *not* consistent with the loss of NegP at stage four.[18]

[16] Although Haeberli (2011) cautions against such a clear-cut distinction.

[17] Van der Wurff (1997, 499) gives the following counterexample in which an NP object moves across *not*, but there are no comparable examples in the PPCME2, so it provides little counter-evidence to the general pattern.

i. And yet they mowe hir lustes nat amende
And yet they can their pleasures not improve
'And yet they cannot raise their spirits'
(c.1390 Chaucer, *Knight's Tale* 3066, van der Wurff (1997, 499, ex.34))

[18] Contracted forms of *not/n't* emerge later. Rissanen (1994, 1999) does not find clear evidence for contracted forms until the sixteenth century. Zwicky and Pullum (1983) argue that these

Table 5.1. *Comparison of the distribution of objects relative to* not *and to adverbs in auxiliated and simplex clauses from the period 1350–1500. Data from the PPCME2.*

| | NP object | | pronoun object | |
|---|---|---|---|---|
| $V_{finite}$ – obj – not – $V_{non-finite}$ | – | (n=0) | 1.3% | (n=4) |
| $V_{finite}$ – not – obj – $V_{non-finite}$ | 100.0% | (n=450) | 98.7% | (n=307) |
| Total | | 450 | | 311 |
| $V_{finite}$ – obj – not | – | (n=0) | 27.8% | (n=153) |
| $V_{finite}$ – not – obj | 100.0% | (n=1005) | 72.2% | (n=397) |
| Total | | 1005 | | 550 |
| $V_{finite}$ – obj – adverb – $V_{non-finite}$ | 1.1% | (n=4) | 2.9% | (n=5) |
| $V_{finite}$ – adverb – obj – $V_{non-finite}$ | 98.9% | (n=344) | 97.1% | (n=165) |
| Total | | 348 | | 170 |
| $V_{finite}$ – obj – adverb – $V_n$on-finite | 38.2% | (n=458) | 66.9% | (n=635) |
| $V_{finite}$ – adverb – obj – $V_n$on-finite | 61.8% | (n=740) | 33.1% | (n=313) |
| Total | | 1198 | | 948 |

### *5.3.4 The Position of* ne: *Evidence from* ne *with Infinitives*

Typically, *ne* is proclitic on finite verbs, which I assume move to $T^0$, or positions higher than $T^0$ throughout Old and Middle English, following Haeberli (2001, 2002a). *ne* appears left-adjacent to the finite verb, irrespective of the position of the verb in $T^0$ (subordinate clauses), $Agr^0$ (uninverted main clauses) or $C^0$ (inverted main clauses). These observations suggest we should associate *ne* [*i*Neg] with finite verbs. This association has often been made, for example by van Kemenade (1999, 152) and Ingham (2007, 390). Assuming that *ne* cliticises on to the finite verb under verb movement, the behaviour of finite verbs negated by *ne* is consistent with a low position for *ne* as the clitic head of a functional projection NegP between TP and vP.

However van Bergen (2012) identifies some environments where *ne* also appears proclitic on a bare infinitive verb.[19] Examples of *ne*+infinitive though rare, are not isolated, random, or the results of scribal errors (writing *ne* in place of *na*), but systematically distributed. Van Bergen describes two contexts in which they appear:

1. Contexts with *uton*: van Bergen (2012, 488) observes this 'combines with a bare infinitive to form a first-person plural adhortative construction comparable to present-day English *let's*' as in (109).

contracted forms of *not* are not syntactic clitics, but inflectional morphemes, posing interesting questions about their syntactic development, and how that development interacts with the development of auxiliaries and the loss of V to T movement during the sixteenth century.

19 van Bergen observes that Old English inflected or *to*-infinitives by contrast are negated by the adverb/particle *na*.

(109) Uton la ne toslitan þa tunecan
let-us lo not asunder-tear the tunic
'Let's not tear asunder the tunic'
(HomS 24 (ScraggVerc 1) 218, Mitchell (1985, §916a))

Van Bergen (2013) claims that there is no evidence for *uton* being anything other than a $C^0$ element. Here it takes a bare infinitive complement.

2. Contexts involving coordination of an affirmative non-finite first conjunct and negative non-finite second conjunct, like (110): in the YCOE corpus, all negative infinitives in this context are negated by *ne*. The only exceptions are either where negation scopes over both first and second conjuncts (in which case the finite verb is negated), or where a negative infinitive appears in negative concord (in which case, the adverbial *na* precedes the non-finite verb).[20]

(110) Hi sculan Godes ege habban on gemynde and ne eargian for
they must God's fear have in mind and not fear for
worldege ealles to swiðe
world-fear entirely too much
'They must keep fear of God in mind and not be afraid of earthly fear at all'
(WPol2.1.2,45, (van Bergen, 2012, 494–5, ex.8))

The issue these examples pose is clearly stated by van Bergen (2012, 512):

> A satisfactory analysis of Old English negation should not predict free attachment of *ne* to non-finite forms; it must account both for the normal restriction to finite verbs and for the possibility of attachment to bare infinitives in the absence of an available finite verb. (van Bergen, 2012, 512)

Assuming that the pre-verbal clitic negative marker is base-generated on the lexical verb $V^0$, as Zeijlstra (2004) does, overgeneralises the distribution of OE *ne*. Examples with the structures in (111) do not occur, despite

[20] The last examples of this construction in the PPCME2 corpus date from c.1225. One, given in i., appears in the *Lambeth Homilies*.

i. and he scal wicche creft aleggan and wiȝgelunge ne gemen
and he shall witch craft put-down and sorcery NEG tolerate
(CMLAMBX1,115.1119)

In the *Vices and Virtues* dating from approximately the same time, we find ii., an non-finite second conjunct negated by *ne...not*.

ii. 'Hu mai ic on ðane world wuniȝen and naht hes ne luuiȝen?'
How may I in the world dwell and not them NEG love?
'How may I dwell in the world and not love them?'
(CMVICES1,41.475)

a high number of negative clauses with both finite and non-finite verbs in the YCOE data.

(111) a. $V_{finite}$ ... *ne*+$V_{infinitive}$
b. (pre-)$modal_{finite}$ ... *ne*+$V_{infinitive}$

The non-occurrence of these patterns is an argument against merging negated verbs in $V^0$ or $v^0$. Van Bergen (2012, 505) observes that the defining characteristic of contexts for *ne*+infinitive is the absence of a finite verb. The issue for us is how to formalise this generalisation and provide a structure for clauses involving *ne*+infinitive.

Van Bergen (2012, 512) provides the key to the issue when she notes that Koopman (2005) observes particularly high frequencies of postverbal pronoun and particle placement in some non-finite clauses. The contexts Koopman identifies for most frequent postverbal placement of particles and pronouns are typically coordinate constructions with a non-finite second conjunct such as (112), or ACI constructions such as (113) – the same contexts where van Bergen finds *ne*+infinitive.

(112) And he sceal forgifan eallum þam mannum, þe him ær
and he must forgive all the men who him before
abulgon, and biddan hym forgifnysse
offended and ask them forgiveness
'and he must forgive all the men who had offended him and ask them for forgiveness'
(ÆLet 3 (Wulfstan 2) 17, (Koopman, 2005, 55, example (30)))

(113) þa het he niman Claudium and lædan to sæ, and wurpan hine
then ordered he take Claudius and lead to sea and throw him
ut mid anum weorcstane
out with a hewn stone
'then he ordered Claudius to be taken out and led to the sea and thrown in with a hewn stone'
(ÆLS (Chrysanthus) 222, (Koopman, 2005, 56, example (33)))

Table 5.2 summarises the figures from Koopman (2005, 59, table 4) for post-verbal particle and pronoun placement in non-finite second conjuncts of the type in (112), ACI constructions of the type in (113), and other non-finite clauses. These data point to leftward movement of the infinitive out of vP in the absence of a finite verb. Thus clauses like (112) involve conjunction of constituents larger than vP. As van Bergen (2012, 512–3, fn.52) notes, without positing leftward verb-movement, it is difficult to explain the higher frequency of post-verbal particles and pronouns in these specific kinds of non-finite clauses. The alternative is to assume a much higher frequency of base VO word order with head-initial VP in just these clauses, but as van Bergen comments, in that case it is difficult to make a principled link between the frequency of VO orders and the specific types of non-finite clauses where post-verbal pronouns and particles are most frequent.

Table 5.2. *Post-verbal particle and adverb placement in non-finite clauses by type, from Koopman. (2005, 59, table 4).*

| | | |
|---|---|---|
| Pronouns | Coordinated<br>20.8% (n=46/221) | Other<br>2.5% (n=81/3125) |
| Particles | ACI<br>37.6% (n=26/69) | Other<br>8.0% (n=26/323) |

Further evidence that these coordinated infinitive constructions involve non-finite verb movement to the head of a functional projection (which may be head-initial or head-final) comes from the examples given in (114) and (115). In (114), the non-finite verb has moved leftwards over the manner adverb *fullice* 'fully', and in (115) it has moved leftwards over the temporal *æfre* 'ever'.[21]

(114) Se mann þe hæfð swynes þeawas, & wyle hine aþwean mid
The man that has pigs' manners and will himself wash with
wope fram synnum, & eft hine befylan fullice mid
tears from sin and again himself defile fully with
leahtrum, swa swa swin deð...
sin, as pigs do...
'The man that has pigs' manners and will wash himself of his sin with tears and then again defile himself fully with sin, as pigs do...'
(coaelhom,ÆHom_ 18:262.2639)

(115) hu men lybban sceoldon, and ænne God wurðian æfre mid
how men live ought, and one God honour ever with
geleafan
faith
'How ought men live and forever honour one God with faith?' (coaelhom,ÆHom_ 21:33.3092)

Taken in combination, these observations suggest that in the absence of a finite verb, non-finite verbs move out of $V^0$ to a higher position, across negation, just as in finite clauses. The distribution of *ne*+infinitive follows if *ne* represents a head $Neg^0$ higher than vP, cliticising onto finite and infinitive verbs under verb movement through $Neg^0$.

A verb movement analysis links the *ne*+infinitive construction to Koopman's postverbal particle constructions, and explains the post-verbal adverb

[21] In many examples where the verb appears to remain in its base $V^0$ position as in i., the verb could equally be analysed as moving to a higher head-final projection.

i. hie his word gehyran moston, & his larum fylgean
They his word hear must, and his teaching follow
'They must hear his words and follow his teaching'
(coblick,LS17.1[MartinMor[BlHom17]]:219.161.2814)

data in (114) and (115). Post-verbal pronouns and particles follow straightforwardly if the pronoun or the particle remains within the VP, but the non-finite verb moves to a higher head-medial functional head position. The landing site of movement seems to be to the left of manner and temporal adverbials, suggesting that it is higher than $v^0$. Roberts (1993) and Nunes (1995) suggest OE bare infinitives have non-finite $T^0$, with a functional structure parallel to finite clauses.[22] In that case, the *ne*+infinitive construction is derived by movement of the non-finite verb through the same low $Neg^0$ position above vP that we proposed for finite clauses on the basis of the position of *not*. *ne* cliticises to the non-finite verb under verb-movement in the same way as for finite verbs. It follows that *ne* only cliticises onto non-finite verbs that move to non-finite $T^0$. *ne* cannot cliticise onto verbs that remain in vP and VP. The non-occurrence of the patterns illustrated in (111) follows because the non-finite verbs in those structures project only vP, rather than non-finite TP.

## 5.4 Agree Relations at Successive Stages of the Jespersen Cycle

Having established the positions of negation in clause structure, I now turn to derivation of the relationships between *ne* and *not* at stages two and three of the Jespersen Cycle. Semantically uninterpretable [uNeg] features must enter into a syntactic relationship with a matching LF-interpretable [iNeg] feature so that they are deleted prior to semantic interpretation at LF. These syntactic configurations should derive the dependencies between negative items we see at each stage of the Jespersen Cycle, and also in negative concord. Like Zeijlstra (2004), I argue that the dependencies between negative items within the Jespersen Cycle and in negative concord are derived by feature-agreement (Agree). However, my hypothesis differs from Zeijlstra's in several crucial respects.

### *5.4.1 The Direction of Agree*

Chomsky (2000) derives long-distance agreement dependencies as follows. The dependency is initiated by a probe, and the target of the agreement is a goal. Hiraiwa (2001, 68) summarises Chomsky's position thus:

> ...uninterpretable features of a probe $\alpha$ and a goal $\beta$ are erased under the structural relation (1), subject to the Matching Condition (2).
>
> (1) AGREE (cf. Chomsky 2000) $\alpha > \beta$
>
> AGREE ($\alpha, \beta$), where $\alpha$ is a probe and $\beta$ is a matching goal, '>' is a c-command relation and uninterpretable features of $\alpha$ and $\beta$ are checked/deleted.

[22] Finite and non-finite $T^0$ will be in complementary distribution, explaining the restriction of *ne*+infinitive to contexts lacking a non-finite verb.

(2) Match (Chomsky 2000:122)
   a. Matching is feature identity.
   b. D(P) is the sister of P.
   c. Locality reduces to 'closest c-command'.

The Probe (C-commanding feature) must have an uninterpretable feature which causes it to enter into a dependency with a matching goal. Chomsky also proposes that goals have uninterpretable features in order that they are visible or active goals for Agree.

Hiraiwa (2001) extends the definition of Agree to include multiple active goals, providing no inactive matching goal intervenes between them, as in (116)

(116) MULTIPLE AGREE as a single simultaneous operation: $\alpha > \beta > \gamma$ (AGREE ($\alpha, \beta, \gamma$), where $\alpha$ is a probe and both $\beta$ and $\gamma$ are matching goals for $\alpha$.)

Three key points emerge from these definitions: first, uninterpretable features are probes C-commanding matching goals; second, goals are active by virtue of having an uninterpretable feature themselves; third, multiple agree is subject to locality, checking all matching features on active goals within a local syntactic domain.

At stage three of the Jespersen Cycle shown in (117), *not* [iNeg] in spec,NegP C-commands *ne* [uNeg] in $Neg^0$. *not* [iNeg] is (externally) merged to check the [uNeg] feature on *ne*. This is unproblematic.

(117)

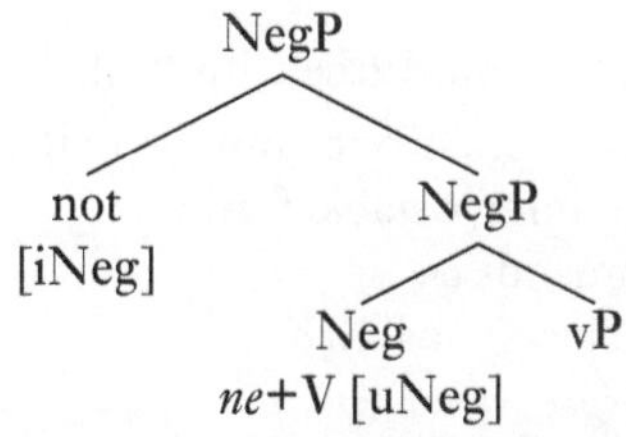


However, analysing the dependency between the negative marker *ne* and the adverbial minimiser *not* at stage two of the Jespersen Cycle under Agree is problematic. At stage two, represented in (118), *ne* is a negative marker with an [iNeg] feature and C-commands the concordant vP-adjoined adverb *not* [uNeg]. Although *not* is an active goal by virtue of its [uNeg] feature, *ne* is not a probe. It has no uninterpretable feature that requires checking.

(118)

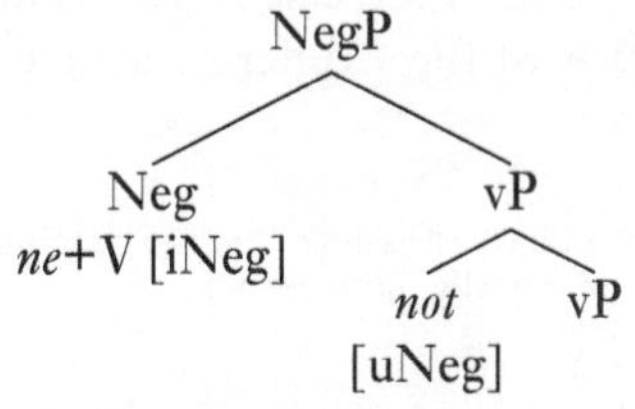

Zeijlstra (2004) posits a structure with a null operator at stage two, as in (119). By making the structure at stage two parallel to the structure at stage three given in (117) Zeiljstra avoids this problem.[23]

(119)

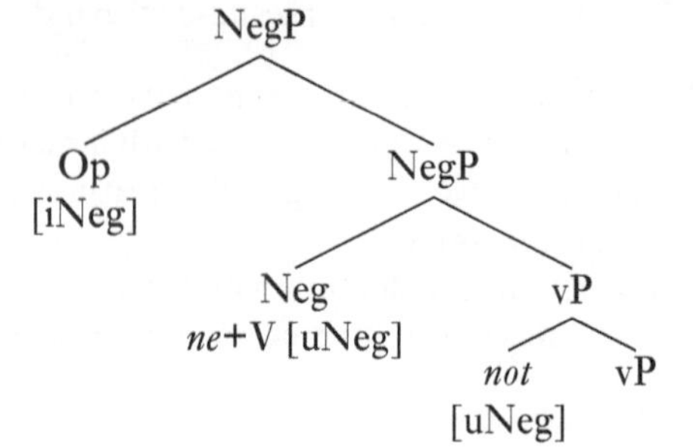


However, this does not allow us to maintain a distinction between stage two *ne* and stage three *ne* in terms of feature interpretability. Consequently, it overgeneralises the distribution of stage two (Old English) *ne*, predicting that it will occur in wide scope negative concord when it does not.

Instead, (118) exemplifies a Reverse Agree configuration of the type proposed by Wurmbrand (2012), Zeijlstra (2012) and Merchant (2013). In Reverse Agree, an uninterpretable or unvalued feature probes upwards for a matching interpretable or valued feature of the same type that C-commands it. In other words, when a lexical item with an [iF] feature is merged, that merge operation will have as a side-effect valuation of all matching [uF] within a local domain. Zeijlstra (2012) defines Reverse Agree in terms of C-command as (120).

(120) Reverse Agree: $\alpha$ can Agree with $\beta$ iff:
  a. $\alpha$ carries at least one uninterpretable feature and $\beta$ carries a matching interpretable feature.
  b. $\beta$ c-commands $\alpha$.
  c. $\beta$ is the closest goal to $\alpha$.

  (Zeijlstra 2012, 17)

Multiple Agree configurations fall out from this definition: multiple probes [uF] can agree with a single goal [iF], providing the interpretable feature C-commands all the uninterpretable features.

In order to maintain that the distinction between stage two and stage three of the Jespersen Cycle is the distinction between (118) and (117) – that the difference between the stages is in the LF-interpretability of the neg-feature on *ne* – we need Reverse Agree. Empirical support for Reverse Agree comes from at least three different syntactic phenomena: VP-ellipsis

[23] Even in this structure, we have to assume that the matching feature on the goal *not* in (119) is not required to be semantically interpretable.

(Merchant, 2013), parasitic participle constructions (Wurmbrand, 2012), negative concord and NPI licensing (Zeijlstra, 2012).[24]

### *5.4.2 Agree and Locality*

Turning now to the locality constraints on Agree, recall that in Middle English biclausal structures like (121), two instances of *ne* may appear in negative concord at stage three of the Jespersen Cycle, one in the subordinate clause and one in the superordinate clause.

(121) ne doute the nat that alle thinges ne ben don aryght
NEG doubt you not that all things ne are don rightfully
'Do not doubt that all things are done rightfully'
(Chaucer's Boethius IV P5,49)

These are biclausal structures, involving non-assertive or subjunctive complement clauses. The embedded clause *ne* is semantically redundant. It is $ne_2$ [uNeg], licensed under agreement with a semantically interpretable negative operator *not*. By treating *ne* in the subordinate clause of (121) as $ne_2$ [*u*Neg] we make two predictions with respect to its distribution, both of which are borne out by the empirical data in Chapter 3: first that it requires licensing by an LF-interpretable negative item, therefore it only appears in wide scope negative concord, and not in contexts where we find non-veridical NPIs (contra Breitbarth (2009)); and second, that the licensing conditions on $ne_2$ are subject to syntactic locality constraints. Negative concord is typically clause-bound, except for contexts like (121). Their exceptionality follows from a phase-based implementation of locality constraints.

In Chomsky (1999), locality is derived by cyclic spell-out of the derivation in phases. CP constitutes a strong phase, therefore only its edge (the head $C^0$ and specifiers or adjuncts of CP) are available to participate in Agree or Move operations within a superordinate phase. This derives successive cyclic movement. It follows that negative concord will typically be clause bound. However, there is evidence to distinguish assertive CPs and non-assertive CPs in terms of phasehood. Giorgi (2004) observes long distance anaphora between a main clause and a non-assertive (or subjunctive) subordinate clause that cannot be explained if the subordinate CP is a phase. In her account, non-assertive CPs are defective, lacking a force specification. The lack of force prevents non-assertive CPs from being strong phases. This predicts wide-scope negative concord is possible only across non-assertive clause boundaries, restricting wide scope negative concord to contexts like (121) where it is attested historically. The derivation of (121) to the point where *not* is merged is given in (122). The complete derivation is given in (123).

[24] Zeijlstra (2012) goes further, arguing that all Agree relations (including successive cyclic movement) can be recast as Reverse (or upward) Agree. However, $\phi$-feature agreement may be an exception (see the counter-arguments to Zeijlstra advanced by Preminger (2013)).

(122)

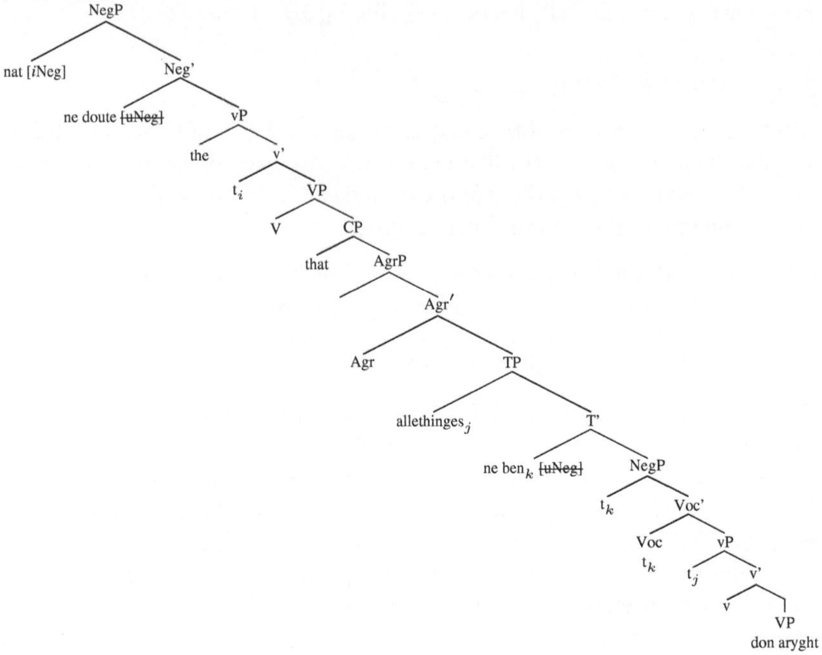
NegP
nat [iNeg]
Neg'
ne doute [uNeg]
vP
the
v'
t_i
VP
V
CP
that
AgrP
Agr'
Agr
TP
allethinges_j
T'
ne ben_k [uNeg]
NegP
t_k
Voc'
Voc
t_k
vP
t_j
v'
v
VP
don aryght

(123)

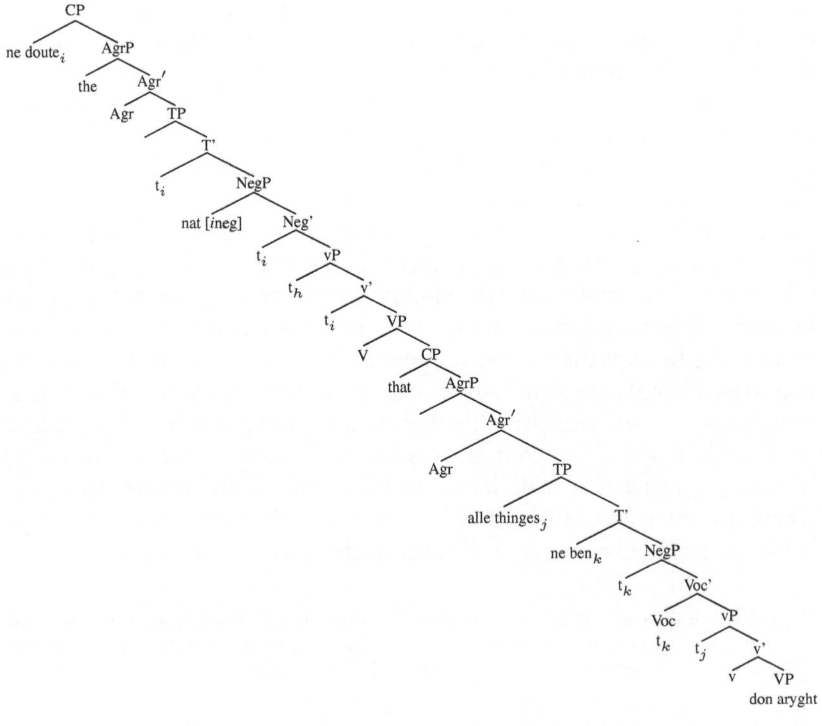
CP
ne doute_i
AgrP
the
Agr'
Agr
TP
T'
t_i
NegP
nat [ineg]
Neg'
t_i
vP
t_h
v'
t_i
VP
V
CP
that
AgrP
Agr'
Agr
TP
alle thinges_j
T'
ne ben_k
NegP
t_k
Voc'
Voc
t_k
vP
t_j
v'
v
VP
don aryght

### 5.4.3 *Negative Inversion*

Finally, I address the derivation of negative inversion in this framework. This requires a slightly more articulated approach to morphosyntactic features, as we must be able to specify which features are to be interpreted at either or both LF and PF interfaces. I assume negative inversion is polarity focus triggered by $C^0$, hence $C^0$ is specified with an [iFocus] feature, and a [uNeg*] feature. The [uNeg*] feature is uninterpretable at LF, but overtly realised at PF, so it triggers overt movement of a negative operator to $C^0$ before spellout. This ensures that a negative operator is focused – associated with $C^0$ at both LF and PF. Adopting an approach similar to Eythorsson (2002), negative inversion follows if the [uNeg*] feature on $C^0$ is satisfied by internal merge of $ne_1$ [iNeg] which head-adjoins to $C^0$ before spellout. This places a negative operator in clause-initial position. Following Roberts and Roussou (2003), the * diacritic on [uNeg*] indicates a feature which must have an overt morphological realisation. If the [uNeg*] feature is checked before spellout, this requirement is met, and the focused clause-initial negative operator is overt. This mechanism can derive polarity focus with a range of clause-initial elements, provided those elements are negative operators with an [iNeg] feature.[25] The category feature C* must also be morphologically overt. Where the [uNeg*] feature is satisfied by head adjunction, C* is already overt. Where the [uNeg*] feature is satisfied by internal or external merge in spec,CP that guarantees only that spec,CP is overt. The additional requirement for C* to be morphologically overt must be satisfied by verb-movement to C at PF, hence subject–verb inversion in these contexts.

As Haeberli (2011) observes, we need to restrict the occurrence of the focus adverb *not* to inverted clauses. The [iFocus] feature licenses [uFocus] in its scope by Reverse Agree (see Section 5.4.1 for discussion of Reverse Agree). Both the function of concordant focus adverb *not* and its restriction to inverted clauses follows if it is specified [uFocus] as well as [uNeg], and only licensed by this particular kind of $C^0$.

## 5.5 Summary

This chapter argued for an analysis of the Jespersen Cycle in which the semantic interpretability of morphosyntactic features associated with *ne* and *not* changes over time. In order to accommodate the empirical observation that there are two forms of *ne* in competition, the account I propose differs in several respects from the accounts of Roberts and Roussou (2003), Zeijlstra

25 If we substitute the [uNeg] feature on $C^0$ with an unvalued affective or operator feature [uAff:__] we can use the same mechanism to derive inversion following affective operators such as *Seldom, Few* in *Seldom have I been so angry*, *Few things have I been so angry about.*

(2004) and Breitbarth (2009). In conclusion, we can maintain a Minimalist account of the early English Jespersen Cycle which accurately predicts the different distributions of *ne* at each stage if we posit the following:

**Features:** There are two forms of *ne* distinguished in terms of the LF-interpretability of their negative features. Analyses based on feature-valuation are difficult to implement in a way that is sufficiently constrained to derive the distribution of ME *ne* as an anti-veridical polarity item. Pesetsky and Torrego (2007) allow features to vary in LF-interpretability and valuation. This provides a potential mechanism to explain the differences between ME *ne* (semantically uninterpretable [uAff:Neg]) and French *ne* (unvalued [uAff:__]). These may be diachronically linked stages, such that a feature first loses its LF-interpretability then its value. English *ne* may simply have been lost before it became unvalued like French *ne*.

**Functional Projections:** The distribution of *ne* and *not* suggest that both occupy low positions in the functional structure of the clause between TP and vP. The position of *not* is distinct from vP-adjoined adverbs, and the position of *ne* distinct from $v^0$. These distributional observations provide evidence for a distinct functional projection NegP. At stage three of the Jespersen Cycle, the head of NegP consists entirely of uninterpretable features. Therefore, we have to admit Feature Scattering (Giorgi and Pianesi, 1997) or Proxy projections (Nash and Rouveret, 1997) to accommodate the distinct position of *not* in spec,NegP at stage three of the Jespersen Cycle.

**Agree:** Agree is required to derive dependencies between negative items. However, the standard 'downwards' agree does not allow us to maintain a distinction between *ne* [iNeg] at stage two of the Jespersen Cycle and *ne* [uNeg] at stage three and formulate the Agree dependencies between *ne* and *not* at each stage. In order to maintain a distinction between *ne* [iNeg] and *ne* [uNeg], Reverse Agree of the type proposed by Wurmbrand (2012), Merchant (2013) and Zeijlstra (2012) is required.

**Locality:** The locality constraints on redundant *ne* at stage three of the Jespersen Cycle follow from a phase-based approach to locality, providing that non-assertive CPs are not phases (as Giorgi (2004) argues).

The analysis has important implications for competition models of change. By allowing long distance dependencies between lexical items, a morphosyntactic feature-based analysis dissociates structural competitors, those items that occupy the same $Neg^0$ position, ($ne_1$ [iNeg] and $ne_2$ [uNeg]); from functional competitors, those items with identical feature specifications, $ne_1$ [iNeg] and *not* [iNeg]. The link between structural and functional competitors becomes an indirect one, satisfied by an Agree relation. This allows different syntactic structures to come into competition with each other at successive stages of the Jespersen Cycle.

Chapters 7 and 8 consider how Reverse Agree derives the negative concord dependencies that are attested in Old and Middle English. Before turning to negative concord, however, Chapter 6 discusses the role of functional change in the Jespersen Cycle. Changes in the function of *ne* during ME provide further support for the distinction between $ne_1$ and $ne_2$.

# 6 The Role of Functional Change in the Jespersen Cycle

## 6.1 Introduction

This chapter examines the role of functional change within the Jespersen Cycle. It investigates whether the diachronic stages identified in Chapters 2–5 perform distinct pragmatic or discourse functions during the periods when they compete, and whether changes in their functions explain changes in the distributions of *ne*, *ne...not* and *not* during the course of the Jespersen Cycle. In particular, I examine how functional change contributes to the grammaticalisation of *not*. Using logistic regression, we can model pathways of functional change within the Jespersen Cycle predicted by different functional analyses. The distributions of negative markers in diachronic corpus data provide an empirical basis against which to test these models, just as they did when we modelled the formal changes within the cycle in Chapter 2.

Within competition models of change, the null hypothesis is that functional change should arise simply as a consequence of morphosyntactic competition, without stipulating any further mechanisms of functional change. Functional constraints should observe the Constant Rate Effect just like the grammatical constraints in Chapter 2 did. Thus the relationship between functional change and the formal model of the Jespersen Cycle as morphosyntactic competition I proposed in Chapter 5 is crucial. I argue once we dissociate functional and structural competitors, in the way a morphosyntactic feature-based analysis allows us to do, functional change arises through competition between forms that are structurally equivalent but functionally distinct.

The idea that functional change plays a role in the grammaticalisation of new negative markers dates back to Gardiner (1904) who says in respect of the French negatives *pas* and *point*:

> These words, from the Latin *passum* and *punctum*, were originally adverbial accusatives placed at the end of negative sentences for the purposes of emphasis, just like the English "not a jot", "not a straw"...*Pas* and *point*...next lose their emphasising force, and become mere adjuncts of the negative words [French *ne*].
> (Gardiner, 1904, 134)

More recently, this idea has been taken up by Detges and Waltereit (2002), Kiparsky and Condoravdi (2006), van der Auwera (2009), Hansen (2009) and Hansen and Visconti (2009). All share in common the basic idea that bipartite negatives such as French *ne...pas* are initially pragmatically marked for particular functions consistent with 'emphasis', and that they gradually become less marked or unmarked over time through processes of functional extension.

Assuming that bipartite negation sets out marked in some way and becomes unmarked over time, how does this functional change relate to the formal changes seen at each stage of the Jespersen Cycle? Are processes of formal and functional change dependent on or independent of each other? Relating Gardiner's comments to the discussion of *not* in Section 4.2, he predicts that the adverbial minimiser *not* (Jespersen Cycle stage two) is pragmatically marked for emphasis, but that it loses this emphatic function when it becomes a negative marker (stage three). This is the simplest hypothesis – that pragmatic unmarking follows from the syntactic reanalysis of *not* as a negative marker. However, van der Auwera (2009: 44) dissociates pragmatic unmarking from the formal changes involved in the Jespersen Cycle. Instead, he hypothesises that the Jespersen Cycle comprises an interaction of formal and semantic strengthening in the marking of negation, in which new sentential negative markers such as English *not* or French *pas* emerge from syntactic reanalysis functionally marked, only later becoming pragmatically unmarked. He schematises the processes for French as in Table 6.1.

Table 6.1. *Van der Auwera's proposed interaction of formal, semantic and pragmatic changes in the Jespersen Cycle (based on van der Auwera (2009, 44, ex.9)).*

| Stage | Emphatic strategies | Neutral strategies | |
|---|---|---|---|
| 1 | | *non*$_{NEG}$ | |
| | | | Formal weakening ↓ |
| 2 | | *ne*$_{NEG}$ | |
| | | | Formal & pragmatic strengthening ↓ |
| 3 | *ne*$_{NEG}$...*pas* | *ne*$_{NEG}$ | |
| | | | Semantic reanalysis of *pas* ↓ |
| 4 | *ne*$_{NEG}$...*pas*$_{NEG}$ | *ne*$_{NEG}$ | |
| | | | Pragmatic weakening of *pas* / Formal strengthening ↓ |
| 5 | | *ne*$_{NEG}$...*pas*$_{NEG}$ | |
| | | | Semantic weakening of *ne* ↓ |
| 6 | | *ne...pas*$_{NEG}$ | |
| | | | Formal weakening of *ne* ↓ |
| 7 | | *pas*$_{NEG}$ | |

Bipartite negation emerges at stage three of this model in contexts pragmatically marked for emphasis. This means that bipartite negation is only possible in a subset of contexts available to neutral negation – a set of contexts that are distinguished by having some additional pragmatic or discourse property. At this point *pas* is an adverbial minimiser in negative concord with the negative marker *ne*. At stage four, *pas* is reanalysed as a negative marker. Only after *pas* is reanalysed as a negative marker, does it become pragmatically unmarked (stage 5), with the effect of formally strengthening the marker of neutral negation. At stage 6, we see the weakening of *ne*, in a similar way to that discussed in Chapters 2–4. However, the loss of *ne* is consequent on pragmatic unmarking of *pas*, so the loss of *ne* will only occur once *pas* is pragmatically unmarked. The key points of this model are:

1. Secondary negators, such as French *pas*, originate as adverbial minimisers performing a particular set of pragmatic functions, in a particular set of linguistic contexts that should be identifiable in corpus data.
2. The reanalysis of the adverbial minimiser as a negative marker takes place in these pragmatically marked environments giving rise to a sentential negative marker which is pragmatically marked.
3. Subsequently, functional extension of this negative marker into new pragmatic contexts causes it to become pragmatically unmarked as the default means of marking sentential negation, at which point the original negative marker *ne* undergoes semantic weakening.

Hansen (2009) and Hansen and Visconti (2009) propose similar models on the basis of medieval French data. Again, they distinguish a pragmatically marked *ne...pas* stage, and hypothesise that pragmatic unmarking leads to a pragmatically neutral *ne...pas* stage. These models are more articulated than the model I developed in Chapter 5. They dissociate processes of semantic, pragmatic and formal strengthening and weakening. The syntactic reanalysis of French *pas* does not affect its pragmatic function – it remains pragmatically marked. All these accounts take data from Romance languages as their starting point. Do such models generalise to other languages? They raise several questions for a morphosyntactic competition analysis of the early English Jespersen Cycle.

1. Is there evidence for a pragmatically marked *ne...not* stage in early English corpus data? If so, how is *ne...not* pragmatically marked?
2. Is there evidence for pragmatic unmarking of *ne...not* in terms of functional extension? – Does *not* spread from a narrowly marked or specialised set of pragmatic contexts to become generalised or unmarked? If so, what trajectory does this functional extension take?
3. Is it possible to analyse pragmatic unmarking within a competition model of language change such as Kroch (1989)?

4. What is the relationship between processes of semantic strengthening and weakening, formal strengthening and weakening, and pragmatic unmarking? Is pragmatic unmarking independent of formal and semantic changes in the way van der Auwera (2009) suggests, or is it related to
   a. the semantic weakening of the original negative marker *ne*1 (competition between *ne*1 and *ne*2), and the consequent semantic reanalysis of *not* from minimiser to negative marker at stage three of the Jespersen Cycle?
   b. the loss of *ne*2 at stage four of the Jespersen Cycle?

## 6.2 Functions of Negation

The first issue to address, before we can investigate pragmatic unmarking, or even whether *not* is pragmatically marked, is what the pragmatic functions of negation are. If a negative marker is pragmatically marked, what does that mean for the functional constraints on its distribution? Israel (1998) explains pragmatic markedness in terms of hyponymy: an emphatic negative marker entails an unemphatic negative marker, but not vice versa. However, linguistic analyses such as Detges and Waltereit (2002), Kiparsky and Condoravdi (2006), Schwenter (2006), Hansen (2009), Hansen and Visconti (2009) characterise the functions of 'emphatic' negatives – and therefore the process of unmarking – differently.

### *6.2.1 Presupposition*

Schwegler (1988, 48) analyses the distinction between emphatic negation and neutral negation in terms of presupposition – emphatic negation is used in contexts where the negation contradicts the presupposed truth value of the proposition. Both Schwegler (1988, 41) and Detges and Waltereit (2002, 183) propose that emphatic negative markers are used to negate a proposition which is already part of the discourse, either stated or implied, and thereby present to the listener's or reader's attention:

> ...markers of emphatic negation are used whenever speakers want to act against some strong counter-expectation on the part of their listeners (Detges and Waltereit, 2002, 183)

The 'counter-expectation' Detges and Waltereit propose describes the cancellation of propositions the listeners presuppose to be true. Detges and Waltereit (2002, 185) further argue that emphatic negation is maximally informative or relevant in a Gricean conversational maxims sense. They hypothesise that functional extension occurs as a result of rhetorically motivated overuse of pragmatically marked negatives. Speakers extend the use of bipartite negation from its original emphatic contexts, in which it is maximally informative, denying the truth of a presupposed assertion, to less informative contexts, in order to emphasise the relevance or informativity of the utterance for purely rhetorical or expressive effect.

### 6.2.2 *Pragmatic Activation*

Schwenter (2006) observes that a definition of emphasis based on presuppositionality does not explain the distribution of bipartite (stage two) negation cross-linguistically. It is too restrictive. For example, Espinal (1993) notes that bipartite *no...pas* in Catalan can be used to agree with a prior negative statement. In these examples, the negated proposition does not counter a belief or expectation within the conversation. In order to capture the distribution of bipartite negation in Romance languages, Schwenter (2006), Hansen (2009) and Hansen and Visconti (2009) distinguish between negated propositions on the basis of their information status, rather than in terms of emphasis or presupposition.

Dryer (1996) and Birner (2006) distinguish different types of proposition according to the relationship between a proposition and its discourse context. The key distinction is whether the information conveyed by a proposition is already given in or known from the preceding discourse, or whether the proposition introduces new information to the discourse. Given information is activated, evoked in the preceding discourse; and accessible, present to the attention of the hearer or reader. New information is not. Discourse-givenness is not the same as presupposition. None of the conversational participants need to believe a proposition for it to be activated – the truth of the proposition does not need to be presupposed. The proposition merely needs to be part of the prior discourse and present to the attention of the discourse participants, irrespective of their beliefs regarding the proposition itself. Therefore, negating an activated or discourse-given proposition does not necessarily counter a listener's or reader's prior belief, expectation or presupposition. An activated, or discourse-given, proposition is simply one that is identifiable from the preceding discourse.

For example, in (124), the negative proposition is activated because it negates a proposition which has already been mentioned in the discourse. It is also emphatic. It cancels the proposition that has just been mentioned, contradicting its truth value.

(124) a. He said he went. **In fact he did not go**
b. He said he had been in the lecture. **In fact he did not attend**

However, in (125), the second negative proposition (in bold face) is activated because the preceding proposition has already brought it to the reader's attention. However, it is not emphatic because it does not contradict the truth value of the preceding proposition. Instead, it repeats or reinforces it.

(125) He said he did not go to the lecture. **He was not in attendance**

While in (124) and (125), a proposition is activated because it has already been mentioned explicitly in the discourse, Birner (2006) observes that this is only one way a proposition can be activated. A proposition is also activated, or discourse-old, if it can be inferred from the preceding discourse. This is illustrated by (126), in which (126b) is inferable from (126a).

(126) a. I'm not hungry
b. I don't want anything to eat

These inferences fall into several groups: logical entailment (126), presupposition (such as the presupposition that is cancelled in the B part of (127)), conversational implicatures (such as the implicature stated in the B part of (128)) and scalar implicatures like the one given in (129).

(127) A: John's sister doesn't live in London – B: John doesn't have a sister.
(128) A: I tried to get the book you wanted – B: I didn't get the book you wanted.
(129) That meal was OK, but it wasn't great.

We can distinguish between emphatic and non-emphatic discourse-old propositions. Discourse-old propositions that cancel inferences or deny propositions, as in examples (124) or (127) are emphatic in the sense of Detges and Waltereit (2002). Discourse-old propositions which preserve or state negative inferences, like (128), or simply repeat an earlier negative proposition, like (125), are not emphatic. Discourse-new propositions are those which cannot be identified from the prior discourse and therefore which are not accessible to the reader from the preceding discourse. They are not previously mentioned in the discourse, or inferentially linked to the preceding discourse. This gives us a five-way distinction between discourse functions, as illustrated in Table 6.2.

Propositions can be subcategorised in two different ways according to their discourse functions: first, according to whether they are discourse-old or discourse-new (Schwenter, 2006, Hansen, 2009, Hansen and Visconti, 2009); and second, whether they are emphatic or not (Schwegler, 1988, Detges and Waltereit, 2002). It is an empirical question whether either of these functional

Table 6.2. *Types of proposition according to emphasis and discourse activation status.*

| Discourse-given | Emphatic | Unemphatic |
|---|---|---|
| | Denial of an antecedent affirmative proposition | Repetition of an antecedent negative proposition |
| | Cancellation of an inference arising from the preceding discourse | Statement of a negative inference arising from the preceding discourse |
| Discourse-new | | |

distinctions correlate with the distributions of *ne*, *ne...not* and *not* during the Middle English period when the three negative forms compete.

Van der Auwera (2009, 47, fn. 6) distinguishes emphatic negation and presuppositional negation, suggesting that, first, pragmatic unmarking would '...take away emphasis, but leave the higher presuppositionality'. It is not clear what the distinction is for van der Auwera, but we can make an empirically well-founded distinction between emphatic negation, in which negation contradicts a proposition presupposed to be true; and discourse-old negation, that is a negation of a proposition present to the reader's attention irrespective of its truth value. Emphasis constitutes a subset of discourse-old or activated contexts for negation. By examining the diachrony of *not* in emphatic and activated contexts, we can see if *not* starts out marking a subset of pragmatic contexts, and how functional constraints on its distribution change over time – whether it spreads from emphatic to activated to unmarked contexts, and how formal and functional changes interact at each stage of the Jespersen Cycle.

## 6.3 Discourse Functional Constraints in the Early English Jespersen Cycle

Schwenter (2006), Hansen (2009) and Hansen and Visconti (2009) show that pragmatic activation – the distinction between discourse-old and discourse-new propositions – is sometimes marked in Romance languages. Schwenter (2006) argues that the choice of Italian *non* (stage one) or *non...mica* (stage two), and Catalan *no* (stage one) or *no...pas* (stage two) is sensitive to the distinction between discourse-old and discourse-new information. He argues in respect of bipartite *no...pas* in Catalan that 'there must be a salient proposition, evoked either linguistically...or situationally for *pas* to be felicitous' (Schwenter, 2006, 333). Hansen and Visconti's (2009) medieval French data show that these types of (discourse-old) propositions account for the majority of occurrences of *mie* and *pas*.

### *6.3.1 Establishing Information Status in Historical Data*

Although historical texts provide only indirect access to the OE and ME writers' semantic and pragmatic competences, by examining the relationship between a proposition and the preceding context in a text, we can identify whether a proposition is activated or not. In the absence of any evidence to the contrary, I assume that the pragmatic competence of an ME speaker would have been based on the same principles as the pragmatic competence of a native speaker of Present-day English. Thus I assume that the same types of inference exist: entailment, conversational and scalar implicature, presupposition, bridging inferences and so on; and that these inferences arise in the same way as in PDE. Hence, when identifying inferences, I use a combination of PDE translations and native speaker pragmatic competence to establish how each individual negative proposition relates to the preceding discourse, seeking to optimise the relevance of that proposition to the coherence of the discourse as a whole. This means that where a plausible inferential link exists between a proposition and the preceding discourse, the proposition is regarded as discourse-given.

Without access to native ME speakers, it is not possible to determine fully the relationship between the text and its context. Some propositions which are not activated through being mentioned or inferred within the text itself may be presupposed because they form part of the reader's wider social or cultural knowledge. This knowledge may be assumed by the writer as part of the context in which their text will be read. It is difficult to identify propositions that are presupposed by the authors to be cultural knowledge or beliefs, since we do not know exactly what constituted this wider cultural or social knowledge, nor can we fully appreciate the nuances of a medieval world view. Thus the relationship between the text and this cultural or social knowledge cannot be considered in the analysis. We cannot make any appeal to an author's or reader's beliefs or assumptions since we cannot establish what they are in a way that is independent of the text itself. Instead, we can only examine the relationship between propositions within the text, and the relationship between propositions in the text and the inferences that pragmatic theory predicts will arise from them: entailment, presupposition, scalar implicature, and conversational implicatures based on maxims of manner, quantity and relevance.

In light of these issues, the Middle English data are coded for discourse function based on a combination of the author's knowledge of Middle English, reference to the Middle English Dictionary (Lewis, 2001) to construct glosses, and use of PDE translations of the corpus texts to check the meaning of propositions within their wider discourse context. On occasion, where the PDE translations are somewhat more free than literal, the coding is based on a literal word for word gloss of the ME text, in preference to the published translation. A very few cases that remained unclear or ambiguous at the end of this procedure were simply excluded from the analysis. One could argue that this procedure results in an analysis of the PDE translations rather than the ME texts themselves. However, the robustness of the results presented in Table 6.3 demonstrate that the discourse functional distinctions made under this analysis correlate in large measure with EME authors' choice of *ne*, *ne...not* or *not*.[1]

[1] The coding for functions was carried out consistently by a single investigator over an uninterrupted period of a few weeks. A portion of the data were then double coded – that is, coded on two separate occasions – to check for inconsistencies in the coding procedure. These were few in number and related to particularly problematic contexts. These particular contexts were checked in the whole dataset and recoded.

Table 6.3. *The distribution of ne, ne...not and not according to discourse function*

| | 1150–1250 | | | | 1250–1350 | | | | 1350–1420 | | | |
|---|---|---|---|---|---|---|---|---|---|---|---|---|
| Function | *ne* | *ne...not* | *not* | Total | *ne* | *ne...not* | *not* | Total | *ne* | *ne...not* | *not* | Total |
| Denial of antecedent p | 27.2% (n=12) | 72.8% (n=32) | – (n=0) | 44 | 10.7% (n=3) | 85.7% (n=24) | 3.6% (n=1) | 28 | – (n=0) | 2.2% (n=1) | 97.8% (n=44) | 45 |
| Repetition of antecedent p | 37.5% (n=3) | 62.5% (n=5) | – (n=0) | 8 | 7.1% (n=1) | 92.9% (n=13) | – (n=0) | 14 | – (n=0) | 15.4% (n=2) | 84.6% (n=13) | 15 |
| Cancellation of inference | 12.8% (n=5) | 84.6% (n=33) | 2.6% (n=1) | 39 | – (n=0) | 89.7% (n=35) | 10.3% (n=4) | 39 | – (n=0) | 10.8% (n=4) | 89.2% (n=33) | 37 |
| Assertion of inference | 17.1% (n=28) | 82.3% (n=135) | 0.6% (n=1) | 164 | 1.9% (n=5) | 88.7% (n=228) | 9.3% (n=24) | 257 | – (n=0) | 15.5% (n=22) | 84.5% (n=120) | 142 |
| **Total Discourse-old** | **18.8% (n=48)** | **80.4% (n=205)** | **0.8% (n=2)** | **255** | **2.6% (n=9)** | **88.7% (n=300)** | **8.6% (n=29)** | **338** | **– (n=0)** | **12.1% (n=29)** | **87.8% (n=210)** | **239** |
| Discourse new | 85.2% (n=335) | 14.5% (n=57) | 0.3% (n=1) | 393 | 39.0% (n=135) | 50.9% (n=176) | 10.1% (n=35) | 346 | 0.7% (n=2) | 9.2% (n=27) | 90.1% (n=265) | 292 |
| Counter-factual | 92.3% (n=24) | 7.7% (n=2) | – (n=0) | 26 | 68.8% (n=11) | 31.2% (n=5) | – (n=0) | 16 | 33.3% (n=4) | 25.0% (n=3) | 41.7% (n=5) | 12 |

### 6.3.2 *Evidence for Pragmatic Activation Constraints on Middle English not at Stage Three of the Jespersen Cycle*

Examples (130)–(134) present Middle English clauses with *ne, ne...not* and *not* in each of the five discourse functions identified in Table 6.2. A trio of examples is given for each discourse function – the (a) examples have *ne*, the (b) examples *ne...not*, and the (c) examples *not*. Bold face highlights the negative markers in each example.

(130) Denial or negation of earlier proposition:

a. þenne þe prest þe menezeð rihtliche teðien. þenne
when the priest them admonishes rightfully tithe then
cumeð þe werse to sume mannes heorte and minzeð hine þat
comes the worst to some men's hearts and advises him that
he swo **ne** do
he so NEG do
'When the priest admonishes them to give their tithes aright, then comes the devil to a man's heart and advises him that he do not so.' (CMTRINIT,215.3014)

b. Alle ðo men ðe swinkeð on ðessere swinkfulle world, alle he
All the men that labour in this toilsome world, all they
swinkeð for sumere hope ðe hie habbeð, ðe hem oft
labour for some hope that they have, that them often
eaten ande beswinkð...Ac ðo ðe swinkeð for ðessere eadi
at end deceives...But those that labour for this blessed
hope, hie **ne** bieð **naht** becaht.
hope, they NEG are **not** deceived
'All the men who labour in this toilsome world, they all labour for some hope they have which often deceives them in the end...But those who labour for this blessed hope, they are not deceived.' (CMVICES1,33.385)

c. For it peyneth hem evere as though they sholde dye anon but
For it pains them ever as though they should die now but
certes, they shal **not** dye
truly they shall **not** die
'For it pains them always as if they should die now, but truly they shall not die'
(CMCTPARS,292.C2.194)

(131) Repetition of an earlier negative proposition:

a.Ariseð þanne ge hauen seten, ac we ne mugen þat don
Arise when you have sat, but we NEG may that do
witðuten his elpe. Seie we þanne to him Domine tu cognouisti
without his help. Say we then to him Domine tu cognouisti
sessionem meam et resurrectionem meam – louerd þu wost
sessionem meam et resurrectionem meam – lord you know

hu ich habbe seten and þat ich **ne** mai wið-uten þin elp
how I have sat and that I **NEG** may without your help
risen.
rise.
'Arise when you have sat, but we are not able to do that without His help. Let us say then unto him – Domine tu cognouisti sessionem meam et resurrectionem meam – Lord, you know how I have sat and that I am unable without your help to rise.'
(CMTRINIT,103.1372)

b. ȝef þu ne cnawest þe seolf…ȝef þu **ne** cnawest **naut**
if you NEG know yourself…if you **NEG** know **not**
þe seolf
yourself
'If you not know yourself (CMANCRIW,II.80.941)…If you not know not yourself (CMANCRIW,II.80.948)'

c. And secoundely, he that is irous and wroth, he ne may nat
And secondly, he that is angry and wrathful, he NEG may not
wel deme and he that may **nat** wel deme may nat wel conseille
well judge and he that may **not** well judge may not well advise
'And secondly, he that is angry and full of wrath, he may not judge well, and he who may not judge well may not give good advice.'
(CMCTMELI,222.C1.195)

(132) Statement of a negative inference:

a. Drihten seið ec on his godspelle þet þa beoð godes bern
Lord says also in his gospel that they are God's children
þe beoð isibsumme ac sake **ne** sturiað.
that are peaceable but strife **NEG** raise-up
'The Lord says also in his gospel that they are God's children who are peaceable and do not raise up strife.'
(CMLAMBX1,113.1092)

b. Ich nam noht giet sad of mine sines and forþi **ne** mai
I NEG-am not yet sated of my sins and therefore **NEG** can
ich hie **noht** forlete.
I them **not** renounce
'I am not yet sated of my sins and therefore I cannot renounce them'
(CMTRINIT,75.1028)

c. þei dreven Brut out of þe lande & wolde **not** suffre hym
tehy drove Brut out of the land and would **not** allow him
among hem.
amongst them
'they drove Brut out of the land and would not allow him amongst them'
(CMBRUT3,6.131)

(133) Cancellation of an inference:

a. And þah þes patriaches alse abel and noe and abraham
And though the patriarchs as Abel and Noah and Abraham
and ysaac gode men weren þurh þet ho weren itende
and Isaac good men were through that they were enlightened
of þan halie gast and al þos godnesse hom **ne** mihte
of the holy ghost yet all this goodness them NEG could
werien þet ho ne wenden alle in to helle.
prevent that they redundant-NEG go all in to hell.
'And though the patriarchs as Abel and Noah, Abraham and Isaac, were good men, being enlightened of the Holy Ghost, yet all this goodness could not preserve them from going into hell…'
(CMLAMB1,81.153)

b. and þe lage hadde þo alle þe mihtes þe haueð nu fulluht
and the law had then all the virtues that has now baptism
for ðat clensede þe man of sinne: swa doð nu fulluht ac it
for that cleansed the man of sin: as does now baptism but it
**ne** openede hem **noht** þe blisse of heuene alse fulcneng doð
NEG opened them **not** the bliss of heaven as baptism does
us.
us.
'And that rite had then all the virtues which baptism now has, for that cleansed man of sin even as baptism now does, but it opened not to them the bliss of heaven as baptism does to us.'
(CMTRINIT,87.1165)

c. Als es o þaim at saie als þe gold þu may se, þat semis
Also is of them that say then the gold you may see, that seems
gode and es **noht**:
good and is **not**:
'Also is of them that say then the gold you may see, that seems good and is not'
(CMBENRUL,4.105)

(134) Discourse-new proposition:

a. and here wuneð on wanrede and þoleð his unwilled,
and here dwells in distress and suffers his discomfort,
hwile druie and hwile wete hwile chele wile
sometimes dry and sometimes wet somtimes cold sometimes
hete hwile hunger wile þurst…hwile unhele
hot sometimes hunger sometimes thirst…sometimes sickness
hwile sorinesse and wile werinesse and hwile
sometimes soreness and sometimes weariness and sometimes
wurmene cheu and fele oðre þe ich telle **ne** mai and ne
worms' biting and many others that I tell NEG can and NEG
mai wiðuten helpe him seluen þerwið werien
may without help himself therewith protect

'and here he dwells in distress and endures discomfort, sometimes dry and sometimes wet, sometimes cold and sometimes hot, sometimes hunger, sometimes thirst, sometimes sickness sometimes soreness, sometimes weariness and sometimes the biting of worms, and many others that I cannot tell of, and may not protect himself against them without help'
(CMTRINIT,123.1648)

b. Ðis sade sanctus Iacobus, ðe hali apostel: "Swiðe michel
This said Saint James the holy apostle: "So much
helpð þas rihtwismannes bede", and ðar of he seið ðelliche
avails the righteous's prayer", and there of he told such
forbisne: "Hit was on mann, dedlich alswa we bieð, and he
example "It was a man mortal as we are, and he
besohte at gode þat **naht ne** scolde reinin, for ðe folc to
besought of God that **not** NEG should rain for the people to
kastin. Godd him ihierde – þat was Helyas – and wiðheld alle
chasten. God him heard – that was Elias – and withheld all
reines þrie hier and six moneþes..."
rains three years and six months..."
'This said St. James, the holy apostle: "The prayer of a righteous man avails much", and he gives such an example of it: "There was a man, mortal as we are. and he besought God that it should not rain, to chasten the people. God heard him – that was Elias – and withheld all the rains three years and six months..."'
(CMVICES1, 143,1787)

c. And moorover, whan oure Lord hadde creat Adam, oure
And moreover, when our Lord had created Adam, our
forme fader, he seyde in this wise: It is **nat** good to been a man
first father, he said in this way: It is **not** good to be a man
alloone...
alone...
'And moreover, when our Lord had created Adam, our father, he said this: "It is not good to be a man alone..."'
(CMCTMELI,221.C2.165)

Gardiner (1904, 134) implies that pragmatic change follows from the syntactic renanalysis of an adverbial minimiser as a negative marker at stage three. Detges and Waltereit (2002) and Hansen (2009), Hansen and Visconti (2009) all take up this position. Section 4.2 argued that the Middle English negative marker *not* develops through syntactic and semantic reanalysis of the Old English adverbial NPI minimiser *not*. I argued that the distribution of Middle English *not* suggests an early date for this reanalysis, prior to the earliest ME texts. So, if pragmatic unmarking is a consequence of the syntactic reanalysis of *not* as a negative marker, we should find that ME *ne...not* is pragmatically unmarked from the twelfth century onwards. On the other hand, van der Auwera (2009) predicts that pragmatic unmarking will follow this syntactic reanalysis. Therefore twelfth- and thirteenth-century *not* may remain pragmatically marked even after its syntactic reanalysis, and pragmatic unmarking may be evident later in the thirteenth and fourteenth centuries.

Table 6.3 reports the frequency of *ne*, *ne...not* and *not* in each of the five discourse functions exemplified in (130)–(134). The data come from the PPCME2 (Kroch and Taylor, 2000), and include all instances of sentential scope negation in finite clauses from the periods 1150–1250, 1250–1350, and a sample of clauses from the period 1350–1420. The data for this period comprise 25 per cent of the negative main clauses and negative finite subordinate clauses from each text.[2] The resulting database is much larger than the ones used in previous studies of the discourse functions of negative clauses, such as Hansen (2009) or Hansen and Visconti (2009).

The distributions of *ne* and *ne...not* are constrained by pragmatic activation (the discourse-new vs. discourse-old distinction), but not by emphasis. In the period 1150–1250, *ne...not* marks discourse-old information while *ne* marks discourse-new information. *ne...not* is significantly more frequent in all discourse-old contexts than in discourse-new contexts: $\chi^2(1df)=281.8$, $p\leq.001$. However, it is not more frequent in emphatic discourse-old contexts than unemphatic discourse-old contexts.[3] The functional distinction between *ne* and *ne...not* is most clear in the period 1150–1250. It is maintained in the period 1250–1350, although the contrast between the two contexts appears to weaken as the frequency of *ne...not* rises in all contexts. Figure 6.1 illustrates that the frequencies of *ne...not* in each discourse-function appear to converge over time, as do the frequencies of *ne*. By the mid fourteenth century, *ne* is almost absent from both discourse-old and discourse-new propositions. Instead, clauses which retain *ne* tend to be counterfactuals, such as (135).[4]

(135) For soothly oure sweete Lord Jhesu Crist hath spared us..., that if
For truly our sweet Lord Jesus Christ has spared us..., that if
he ne hadde pitee of mannes soule, a sory song we myghten alle
he NEG had pity of man's soul, a sorry song we might all
synge.
sing

[2] The rationale for this is simply to provide a sample of around 500 clauses for pragmatic analysis, a sample size that is large enough to give reliable statistical results, but not so large as to be impractical to code for pragmatic function, given how time consuming the process of coding the data for pragmatic function is.

[3] The figures for *ne...not* in these two contexts are as follows: emphatic – denial of an antecedent proposition and cancellation of an inference 79.2% (n=65/82); unemphatic – repetition of an antecedent negative proposition and statement of an inference 82.6% (n=38/46). $\chi^2(1df)=.209$, $p=.65$.

[4] Hansen (2009) also identified counterfactuals as a context which retains *ne* in medieval French.

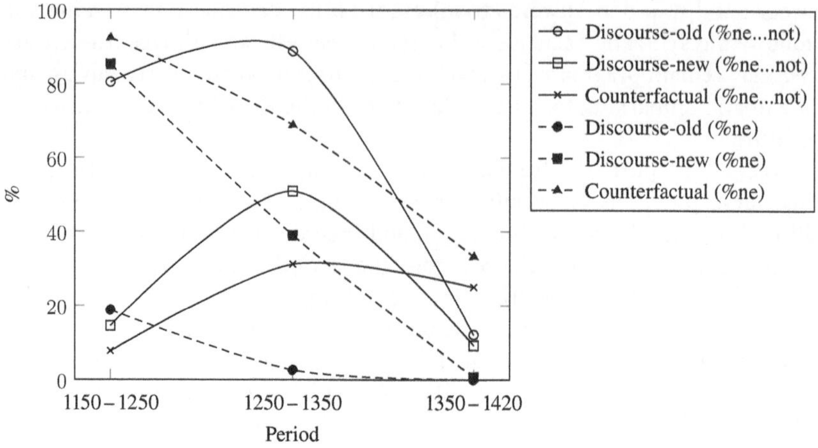


Figure 6.1 The frequencies of *ne* and *ne...not* in discourse-old, discourse-new and counterfactual propositions.

'For truly, our sweet Lord Jesus Christ has spared us...that if he didn't have pity on man's soul, a sorry song we might all sing' (CMCTPARS,296.C2a.344)

Given the pre-twelfth century date I hypothesised for the reanalysis of *not* as a negative marker in Section 4.2, the fact that Middle English *not* remains highly pragmatically marked in the period 1250–1350 indicates that there is no correlation between the syntactic reanalysis of *not* and pragmatic unmarking. Pragmatic function continues to constrain the distribution of ME *not* long after it ceases to be an NPI adverbial minimiser and becomes a negative marker. The syntactic and semantic reanalysis of *not* as a negative marker in the early ME *ne...not* construction is not accompanied by pragmatic unmarking. Instead, a functional distinction between clauses with *ne* and *ne...not* exists throughout ME, but it appears to be different in early ME (discourse-old versus discourse-new) and late ME (counterfactual versus factual). The contexts for *ne* in late ME are much more pragmatically restricted than those available to *ne* in earlier ME.

### 6.3.3 *Evidence for Pragmatic Unmarking at Stage Three of the Jespersen Cycle?*

The accounts proposed by Detges and Waltereit (2002), Hansen (2009) and Hansen and Visconti (2009) all argue that there is a link between pragmatic unmarking of bipartite negative markers, specifically French *ne...pas*, and their increasing frequency over time. They propose various mechanisms to explain this link. Detges and Waltereit (2002) argue that pragmatic unmarking of *pas* results from the pragmatically motivated overuse of *pas* for rhetorical effect. Since *pas* is initially emphatic and emphatic contexts are highly informative and relevant to the discourse, speakers extend the use of *pas* to pragmatically unmarked contexts in order to emphasise the relevance or informativeness of their contribution to the discourse, even when their contribution is not especially relevant or informative. Once *ne...pas* is routinised in contexts that are not especially relevant or informative, *pas* loses its emphatic function and hence becomes grammaticalised as a pragmatically neutral negative marker. Kiparsky and Condoravdi (2006, 5) appeal to a similar 'inflationary effect' in the development of Greek negation stating that:

> Emphatic negation tends to increase in frequency due to pragmatically motivated overuse which is characteristic of inherently bounded evaluative scales. This rise in frequency at the expense of plain negation has an inflationary effect, well attested also in politeness systems, hypocristics, pejoratives and scalar adjectives of all kinds...
> (Kiparsky and Condoravdi, 2006, 5).

This inflationary effect predicts that the pragmatic constraints on emphatic negative markers will weaken over time, causing their extension into new pragmatic functions as they go from pragmatically marked to unmarked. However, an alternative hypothesis – not discussed by Detges and Waltereit (2002), Kiparsky and Condoravdi (2006) or Hansen and Visconti (2009), but suggested by Schwegler (1988) – is that loss of the contrast between discourse-old and discourse-new propositions in French simply results from the increasing frequency of *ne...pas* overall, without any weakening of the pragmatic constraints themselves.

Which of these hypotheses can account for the distribution of Middle English *ne...not* in Table 6.3? Frequency data are insufficient to distinguish between these two hypotheses. Kroch (1989) argues that because change follows a logistic curve, it is not possible to determine from frequency data alone whether the contextual constraints on an innovative form change as it becomes more frequent. The frequencies of a form such as *ne...not* in discourse-given and discourse-new propositions could appear to converge as the frequency of *ne...not* increases, without there being any change in the underlying discourse-functional constraints on the use of *ne...not*, or any difference in the rate at which *ne...not* is introduced in the two contexts. Instead, differences in the frequencies of *ne...not* in the two contexts might arise simply because the starting frequency of *ne...not* is higher in discourse-given propositions than in discourse-new ones.

Figure 6.2 shows the relationship between two logistic curves whose slope (rate of change) is identical, but whose intercepts (the point at which the curve

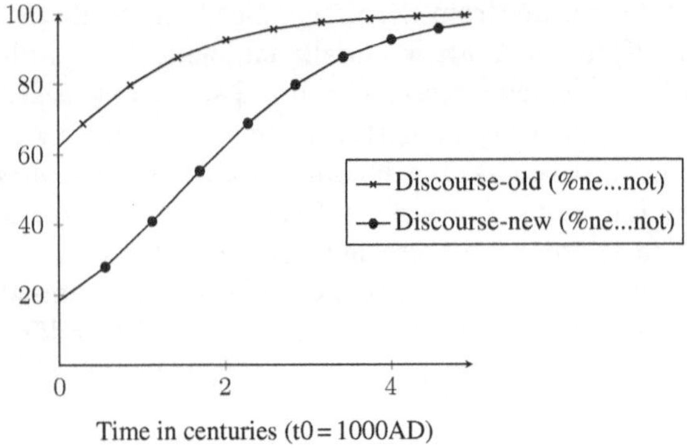


Figure 6.2 Parallel logistic curves in discourse-old and discourse-new functions.

intersects the Y-axis – the starting frequencies of *ne...not*) are different.[5] This provides an illustration of how the frequencies of *ne...not* in the two contexts appear to converge over time, in a logistic model where the contextual constraints on *ne...not* and the rate at which it is introduced both remain constant over time. The frequency of *ne...not* appears to increase more in discourse-new propositions where the starting frequency of *ne...not* is lower. The change in this context starts out near the bottom of the curve, where the frequency of the innovative form is accelerating most. In discourse-old propositions where the starting frequency of *ne...not* is higher, the frequency of *ne...not* is nearer the top of the logistic curve, where frequency of the innovative form is slowing down. Hence it appears that the frequency of *ne...not* increases more rapidly in discourse-new propositions, even though the rate of increase in both contexts – the slope parameter of the logistic curves – is exactly the same.

It also follows that the relationship between the two curves in Figure 6.2 is the same throughout the entire change. This relationship between the curves in each of the contexts for *ne...not* constitutes the contextual constraints on the use of *ne...not*, and these are what we need to determine to identify pragmatic unmarking. If the change proceeds at the same rate in each context, then as the rate of change is independent of (i.e. the same in) the two

[5] The curves are produced by the function

$$\text{p}\,(not) = \left(\frac{e^{k-st}}{1+e^{k-st}}\right) \text{ (Kroch, 1989, 202, ex.1)}$$

where $e$ is a constant, Euler's number, 2.718; $t$ represents time; $k$ represents the intercept parameter, the frequency of the advancing form at the beginning of the change, time t=0; and $s$ is the slope of the curve. The only difference between the two curves in Figure 6.2 is in the intercept parameter $k$ which is higher in discourse-old contexts than in discourse-new ones.

contexts, the relationship between the contexts (the contextual constraints) will remain the same throughout the change in probabilistic terms. Functional constraints are estimable at successive points in the change using a series of logistic binomial regression (or multivariate) analyses, allowing us to compare the constraints within datasets from different time periods. Kroch explains:

> A constant rate of change across contexts is mathematically equivalent to fixity of contextual effects, in direction and size, across time periods. Thus, if a study reports a series of multivariate analyses for different time periods, and the contextual effects are constant across these analyses, the rate of change of each context measured separately would necessarily be the same. This equivalence holds because, in statistical terms, the constant rate hypothesis is the claim that the overall rate of use of a form is independent of the contextual effects on its use.
> (Kroch, 1989, 204)

Each regression (or multivariate) analysis will report, for a given overall frequency of *ne...not* at a particular point in time, how much more likely discourse-old contexts are than discourse-new contexts to employ *ne...not* rather than *ne*. As these probabilities in marked (discourse-old) contexts are calculated relative to discourse-new contexts (the reference level) in each period, they allow us to estimate contextual constraints while taking into account that the overall frequency of *ne...not* increases along a logistic curve over time. Performing these kind of regression analyses allows us to distinguish two possible pathways for pragmatic constraints within a model of competition between *ne* and *ne...not*:

**Pragmatic Unmarking** If *ne...not* undergoes pragmatic unmarking, pragmatic unmarking will manifest itself as a weakening in the discourse-functional constraints on *ne...not* over time, so that:

a. *ne...not* is more likely to occur in discourse-new contexts (the pragmatically unmarked context for *ne...not*) at later points within the change when it is more frequent, than it does at earlier points within the change when it is less frequent. The distinction between discourse-given and discourse-new contexts will weaken, in probabilistic terms, over time. Therefore, the rate of change will be faster in unmarked than marked contexts.
b. the effect of the discourse-given/discourse-new distinction on the distribution of *ne...not* becomes weaker over time in terms of the level of significance this constraint achieves in a model of the variation between *ne* and *ne...not*.

**No pragmatic unmarking** Conversely, if pragmatic unmarking does not take place, then the constraints on *ne* and *ne...not* will remain the same in probabilistic terms, irrespective of the increasing overall frequency of *ne...not* over *ne*.

a. The pragmatic constraints on *ne...not* will only be lost when its use becomes categorical (100 per cent in all contexts), at which point it will be grammaticalised as the default marker of sentential negation in all contexts irrespective of discourse function. The rate of change will be the same in all discourse functions.
b. The Constant Rate Effect is the null hypothesis within a grammar competition model – the scenario follows from a model in which the pragmatic constraints on *ne...not* are independent of its overall frequency of use. Therefore there is no need to stipulate any additional mechanisms of pragmatic unmarking as the frequency of *ne...not* increases.

Table 6.4 reports the results of two separate mixed-effects regression analyses, one for 1150–1250, the other for 1250–1350. The data for each regression analysis come from the respective columns of Table 6.3. The analyses model the likelihood of finding *ne...not* in each discourse function rather than *ne*.[6] They include discourse-function as a fixed effect (predictor) with the five levels given in Table 6.3. Corpus text is incorporated as a random effect, in order to incorporate differences between individual texts into the model.[7]

The models estimate the effects of each discourse-function on the probability of finding *ne...not* rather than *ne* using treatment coding. The probabilities estimated by the models are estimates of how much more likely (in log-odds) *ne...not* is to occur in each discourse-old context rather than in discourse-new propositions. Discourse-new propositions constitute the reference level. This amounts to operationalising the assumption that discourse-new propositions are pragmatically unmarked, and that the analysis will establish the degree to which the other discourse-functions are marked relative to discourse-new propositions.

The regression performs a logistic transform on the frequency data, taking into account that change follows a logistic function (S-curve) over time. Thus we can compare the constraints at successive periods by comparing the estimated probabilities in log-odds. Positive log-odds indicate that *ne...not* is favoured in that context over discourse-new propositions. The higher the log-odds the greater the favouring effect. The p-value associated with each context indicates whether the likelihood of *ne...not* occurring in that context is significantly different from its likelihood of occurrence in discourse-new propositions.

6 Setting aside *not* which occurs independently of *ne* for the moment. I return to the discussion of *not* in Section 6.4.
7 See the discussion of mixed-effects regression models in Section 2.3.2 for more detailed description of these models and the reasons for adopting them for historical corpus data.

Table 6.4. *Logistic regression analysis of factors involved in competition between ne...not (the application value) and ne.*

| | 1150–1250 | | 1250–1350 | |
|---|---|---|---|---|
| Overall frequency *ne...not* (n) | 39.2% (n=264/674) | | 68.2% (n=481/705) | |
| Discourse-function | Effect | p | Effect | p |
| Denial of antecedent p | 2.78 | .001 | 1.92 | .018 |
| Repetition of antecedent p | 2.62 | .001 | – | – |
| Statement of inference | 3.38 | .001 | 3.34 | .001 |
| Cancellation of inference | 3.43 | .001 | – | – |

Some contexts have too few tokens of *ne...not* (<4 tokens of *ne* or of *ne...not*) for reliable estimation of a factor weight within the regression analysis. These contexts are excluded from the analysis and are absent from Table 6.4. Despite the gaps in Table 6.4, we see that the effect of discourse function on competition between *ne...not* and *ne* is consistent across both periods in contexts where there are sufficient data for regression analysis to be performed. This is in spite of a large increase (29.0 per cent) in the overall frequency of *ne...not* in the dataset as a whole. Particularly obvious is the consistency and scale of the difference between the likelihood of *ne...not* in propositions inferentially linked to the preceding discourse versus discourse-new propositions.

By changing the reference level (relevelling) in the analysis to one of the activated discourse-functions, such as 'denial of an antecedent proposition' we see that distributions of *ne...not* in the four activated discourse contexts in Table 6.4 are not significantly different from each other, but that discourse-new propositions differ significantly from all the discourse-old propositions. This suggests we can reduce the analysis from five contexts to a simple two way opposition between discourse-old (activated) and discourse-new (non-activated) propositions.

These results demonstrate that discourse functional constraints remain constant across the two periods, and independent of changes in the overall frequency of *ne...not* versus *ne*.[8] These data provide no evidence of competition between pragmatically marked and pragmatically unmarked *ne...not* stages over time, nor does there appear to be a shift in the pragmatic constraints on *not* from emphasis to pragmatic activation during ME. Instead, the pragmatic constraints on competition between *ne...not* and *ne* are stable from the twelfth to fourteenth centuries. Fitting a logistic curve to the change

[8] One potential explanation might be that the texts in each period are similar in date, clustering just before and after 1250. However, this is not the case. The period 1150–1250 includes many texts from c.1225, but the majority of the texts in the period 1250–1350 date from the early to mid fourteenth century, so c.1325–50. Therefore the effect of these pragmatic constraints does appear to be consistent for over a century at least.

across the period 1150–1350 estimates a rate of change of 2.1 logits/century, with no statistically significant ($p \leq .05$) interactions between the date of each text and the discourse-functional constraints. This confirms that these functional constraints are independent of the changing overall frequencies of *ne* and *ne...not*.

## 6.4 Discourse-functional Constraints, Pragmatic Unmarking and Evidence for Two Forms of *ne*

The functional constraints on $ne_1$ and $ne_2$ differ, providing further evidence to distinguish the two forms. If we analyse *ne* as a single lexical item at stages one, two and three of the Jespersen Cycle, Table 6.5 shows that the functional constraints on *ne* change over time as *ne* is lost. *ne* goes from being significantly disfavoured in discourse-old propositions to being functionally unmarked.[9] The loss of *ne* is not independent of its discourse function. However, when we break down the loss of *ne* into two processes – loss of stage one/two $ne_1$ and loss of stage three $ne_2$ – the functional constraints on each form of *ne* differ from each other, yet are consistent over time in accordance with the Constant Rate Effect.

Table 6.4 showed a clear functional contrast between $ne_1$ (favoured in discourse-new propositions) and $ne_2$*...not* (favoured in discourse-old propositions) in the first ME period (1150–1250). This contrast is maintained in the period 1250–1350, although here there is a more general contrast between $ne_1$ and the two negation strategies involving *not* ($ne_2$*...not* and *not*). The effects of discourse-function on competition between $ne_1$ and $ne_2$*...not* and $ne_1$ and *not* in the periods 1250–1350 are parallel. Both *ne...not* and *not* are equally favoured in discourse-old propositions over discourse-new ones, until $ne_1$ is lost. In the period 1250–1350:

**$ne_1$ vs. $ne_2$*...not*:** disfavouring effect of discourse-old contexts on $ne_1$ (in logits) = -2.78, p=.001

**$ne_1$ vs. *not*:** disfavouring effect of discourse-old contexts on $ne_1$ (in logits) = -2.60, p=.001

Table 6.5. *The effect of the discourse-old/discourse-new contrast on the overall distribution of ne.*

| Date | Effect | p |
|---|---|---|
| 1250–1350 | -0.52 | .06 |
| 1350–1420 | 0.28 | .35 |

[9] Although there is a switch to discourse-old propositions slightly favouring *ne* in the period 1350–1420, any difference in the distributions of *ne* in discourse-new and discourse-old propositions is not statistically significant in this period.

Table 6.6. *Logistic regression analysis of pragmatic constraints on ne...not and not.*

| | 1250–1350 | | 1350–1420 | |
|---|---|---|---|---|
| Frequency at which *ne* co-occurs with *not* (n) | 88.1% (n=476/540) | | 10.4% (n=56/536) | |
| Discourse-function | Effect | p | Effect | p |
| Discourse-old | 0.62 | .02 | 0.43 | .17 |

This is strong evidence that competition between $ne_1$ and *not* and between $ne_1$ and $ne_2$...*not* is in fact a single process subject to the same functional constraints – competition between $ne_1$ [iNeg] and *not* [iNeg]. While ever and wherever these two featurally equivalent lexical items compete, that competition is subject to pragmatic constraints. The negative marker *not* remains pragmatically marked at stages three and four of the Jespersen Cycle until $ne_1$ is lost. Pragmatic unmarking of *not* results from the loss of its competitor $ne_1$, rather than from the loss of $ne_2$ or from any change to *not*. Loss of $ne_1$ creates pressure to generalise *not* across all pragmatic contexts to meet the need to mark clausal negation. This follows straightforwardly in a competition model of change without the need for any additional mechanism of pragmatic unmarking. *not* remains pragmatically marked until it becomes the grammatical default negative marker and its use is categorical.

Turning to $ne_2$, the competition between $ne_2$...*not* and *not* reported in Table 6.6 reveals functional constraints on $ne_2$ which pattern in the opposite direction to the constraints on $ne_1$. In the period 1250–1350, there is a small but significant difference between discourse-old and discourse-new propositions. Discourse-old propositions favour $ne_2$...*not* over *not*. The same effect persists into the period 1350–1420 but it is not statistically significant at the $p<.05$ level due to the smaller number of $ne_2$...*not* tokens in this later period.[10]

The different functions associated with *ne* at successive stages of the Jespersen Cycle indicate that $ne_1$ (stage one) and loss of $ne_2$ (stage two) are functionally as well as syntactically independent of each other. $ne_1$ is highly favoured in discourse-new propositions, whereas $ne_2$ is slightly favoured in discourse-old propositions. The probabilistic effect of function on the distribution of both *ne* and *ne...not* diminishes in significance as the two forms of *ne* both become less frequent in the data, but the functional constraints do not entirely disappear until both forms of *ne* are lost in the fourteenth and fifteenth centuries. The losses of $ne_1$ and $ne_2$ eliminate competition in sentential negative strategies, resulting in the generalisation of *not* across all sentential negation contexts irrespective of their discourse function.

[10] Table 6.6 only includes texts which exhibit variability between *ne...not* and *not*. Those from which *ne...not* is categorically absent are excluded.

Table 6.7. *The effect of the discourse-old/discourse-new contrast on the distributions of ne ne...not and not.*

| | *ne* | | *ne...not* | | *not* | |
|---|---|---|---|---|---|---|
| Date | Effect | p | Effect | p | Effect | p |
| 1150–1250 | -3.23 | .001 | 3.01 | .001 | – | – |
| 1250–1350 | -2.64 | .001 | 1.74 | .001 | 0.52 | .06 |
| 1350–1420 | – | – | 0.43 | .17 | -0.28 | .35 |

Table 6.7 demonstrates how the pragmatic constraints on each of the forms *ne*, *ne...not* and *not* interact over time when all three forms are analysed together. The table shows the extent to which discourse-old contexts favour or disfavour each negative marker over discourse-new contexts. Positive effects indicate that a negative marker is favoured in discourse-old contexts over discourse-new ones, negative effects that a negative marker is disfavoured in discourse-old contexts over discourse-new ones.

The functional contrasts between the three forms diminish as *ne*1 and *ne*2 are lost. There are two intersecting processes of competition here, which I have demonstrated are each subject to stable pragmatic constraints – *ne*1 vs. *not* and *ne*2 vs. *not*. However, in combination, the intersection of these two processes gives rise to the apparent pragmatic unmarking of *ne*2*...not* during ME. This is an epiphenomenon, arising from different functional constraints as *ne*2*...not* first competes with *ne*1, then with *not*. In a model of competition which structures the changes involved in the Jespersen Cycle as I have here, pragmatic unmarking falls out of stable constraints on competition between forms over time rather than requiring extra mechanisms to explain it. Crucially, structuring the competition between forms involves making a distinction between *ne*1 and *ne*2. The losses of *ne*1 and *ne*2 are key to the pragmatic unmarking of *not* because they result in the functional extension of stage three *not* into contexts where negation was formerly marked by *ne* alone or by *ne...not*, but these two losses follow different time courses and are subject to different pragmatic constraints.[11]

## 6.5 Implications: 'priming effects' and Functional Constraints

### *6.5.1 Ecay and Tamminga's Analysis*

Ecay and Tamminga (2016) argue that a priming effect constrains variation between negation strategies in EME. The negative marker used in a

[11] In clauses that have negative quantifiers, like *no, nothing* or negative adverbs like *never*, the use of *not* remains variable throughout Middle English and Early Modern English, and provides the context for Present-day English variation between no-negation and not-negation (Tottie, 1991a). In these contexts, *not* remains pragmatically marked even in PDE. See Section 6.6 for discussion.

Table 6.8. *Comparison of priming effects in models which separate and aggregate discourse-old and discourse-new contexts, period 1150–1250.*

| | Discourse functions separated Target | | | | Discourse-functions aggregated Target | | | |
|---|---|---|---|---|---|---|---|---|
| Prime | ne | ne...not | not | Total | ne | ne...not | not | Total |
| *ne* | 79.9% (n=319) | 19.8% (n=79) | 0.3% (n=1) | 399 | 68.3% (n=269) | 31.3% (n=123) | 0.4% (n=2) | 394 |
| *ne...not* | 30.9% (n=79) | 68.4% (n=175) | 0.8% (n=2) | 256 | 48.1% (n=124) | 51.6% (n=133) | 0.4% (n=1) | 258 |
| *not* | 33.3% (n=1) | 66.7% (n=2) | – (n=0) | 3 | 66.7% (n=2) | 33.3% (n=1) | – (n=0) | 3 |

particular clause (*ne, ne...not* or *not*) is more likely to be the same as the negative marker used in the immediately preceding negative clause than it is to differ from it. A clause negated by *ne* is more likely to follow a clause negated by *ne*, and the same for *ne...not* and *not*. The analysis I propose in Wallage (2008) and here, in Chapter 5, predicts that $ne_1$ and $ne_2$ in *ne...not* are formally distinct. Therefore Ecay and Tamminga argue that priming can be used as a diagnostic to test the fit of my model, and compare it with Frisch's (1997) model. Their argument goes as follows: under Wallage's (2008) model *ne* and *ne...not* involve syntactically distinct forms of *ne*. Therefore, $ne_1$ primes $ne_1$ and $ne_2$*...not* primes $ne_2$*...not*; whereas under Frisch's (1997) account, there is only one form of *ne*, so *ne* and *ne...not* should prime each other.

While Ecay and Tamminga claim their analysis supports the conclusions I drew in Chapters 2–5, their model ignores the functional distinctions between *ne, ne...not* and *not* we saw in this chapter. Tables 6.8 and 6.9 show the fit of the priming model is better when discourse-functions are modelled as separate contexts for priming than when discourse-functions are aggregated (shaded cells are those where there is a priming effect). This simply follows from what we already know: the stronger priming effect when discourse contexts are separated simply reflects the consistency of the pragmatic distinction between *ne* in discourse-new propositions and *ne...not* in discourse-old propositions.

Despite the apparent good fit[12] of the priming model in Table 6.8, here I argue that the priming effect identified by Ecay and Tamminga is not an independent effect, but follows as a consequence of the functionally more articulated model of the Jespersen Cycle outlined in this chapter. Once we

[12] Priming predicts the form of negation in significantly more discourse-new contexts than discourse-old ones (ch-sq(1df)=20.1, p<.001). This follows because negation strategies in discourse-old contexts are more hetrogeneous than in discourse-new contexts, which are more homogenous during this period.

Table 6.9. *Comparison of priming effects in models which separate and aggregate discourse-old and discourse-new contexts, period 1250–1350.*

| | Discourse functions separated Target | | | | Discourse-functions aggregated Target | | | |
|---|---|---|---|---|---|---|---|---|
| Prime | ne | ne...not | not | Total | ne | ne...not | not | Total |
| *ne* | 59.4% (n=94) | 37.9% (n=60) | 2.5% (n=4) | 158 | 45.6% (n=72) | 51.9% (n=82) | 2.5% (n=4) | 158 |
| *ne...not* | 12.3% (n=59) | 78.3% (n=375) | 9.6% (n=46) | 480 | 16.9% (n=81) | 74.4% (n=358) | 8.5% (n=41) | 480 |
| *not* | 7.8% (n=5) | 70.3% (n=45) | 21.9% (n=14) | 64 | 6.3% (n=4) | 65.6% (n=42) | 28.1% (n=18) | 64 |

take functional distinctions between *ne*, *ne...not* and *not* into account, there is no priming effect. This result strengthens evidence for the structural and functional model of the Jespersen Cycle I propose.

### 6.5.2 *Reappraising the Evidence for a Priming Effect*

Under the measure of priming implemented by Ecay & Tamminga, the ideal is a homogenous dataset, in which there is one negative marker and no variation. The closer datasets come to this ideal, the stronger the priming effect will be – not for any psycholinguistic or cognitive reason, but simply as an artefact of the way the data are distributed. Priming will appear to have a stronger effect when datasets are more homogenous and a weaker effect when they are more heterogenous. Datasets will be more homogenous at the early and later stages of a change, rather than in the midpoint of the change. Thus the effect of priming will vary over time as the overall frequencies of forms change. We see this in Tables 6.8 and 6.9: the priming effect is stronger at the beginning of the change in Table 6.8 than as the change progresses in Table 6.9. Furthermore, it will always be the most frequent form at any point in the change that appears to be primed.

This is not priming per se, merely a function of the relative frequencies of forms within the dataset. What we need to establish is whether there is a priming effect which is additional to the changing overall distributions of *ne*, *ne...not* and *not* in the dataset, or whether the priming effect is simply an artefact of the competition between these forms. If there is priming, particular negative markers will cluster together in the dataset such that the likelihood of finding a particular negative marker (NM) in an adjacent pair of clauses is higher than the likelihood of finding that negative marker in two randomly selected negative clauses.

In a dataset where negative markers are distributed randomly, we can calculate the likelihood that a pair of clauses will contain the same negative form as in (136):

(136) Probability that NM will occur in the prime x Probability that NM will occur in the target

If the choice of negative marker in the target clause is independent of the choice of negative marker in the prime, then the likelihood of a particular negative marker appearing in each clause is simply the overall frequency of that negative marker in the dataset as a whole. Thus the likelihood of both clauses involving the same negative maker is the square of the probability of that negative marker occurring in the dataset overall. To demonstrate a priming effect, the likelihood of an adjacent pair of negative clauses having the same negative marker must be significantly higher than the likelihood of a randomly selected pair of negative clauses having the same negative marker.

In the period 1150–1250, *ne* occurs in 60.3% (n=405/672) of clauses. So, the probability that a pair of clauses will both contain *ne* is 36.3%. For *ne...not*, the overall frequency is 39.2% (n=264/672), so the probability that a pair of clauses will both contain *ne...not* is 15.4%. Therefore, the negative marker used in 51.7% of negative clauses will be the same as in the preceding negative clause, even if *ne* and *ne...not* are distributed randomly within the dataset. In fact, the negative marker in 61.4% (n=402/655) of clauses is the same as the preceding negative marker. This is significantly different from the 51.7% (n=338/655) predicted: chi-sq(1df)=12.7, p<.001.

However, when we separate out the data according to pragmatic context – activated vs. non-activated – this evidence for a priming effect disappears, and is merely a function of the different distributions of *ne* and *ne...not* across different pragmatic contexts. Coding for priming within each discourse-context separately – that is, treating only discourse-old antecedents as primes for discourse-old propositions and only discourse-new antecedents as primes for discourse-new propositions – we see that there is no priming effect. The figures are given in Table 6.10.

Overall we predict 76.5% (n=311/407) discourse-new propositions have the same NM as any other negative clause selected from the dataset at random. We observe that in fact 310 clauses with *ne* and 15 clauses with *ne...not* have the same NM as the preceding clause, giving a total of 79.8% (n=325/407). This observed total is not significantly different ($\chi^2$(1df)=1.41, p=.235) from the total predicted if there is no priming effect. If we break down the data in Table 6.10, we find no significant differences between the predicted and observed frequencies of priming in discourse-new propositions involving either *ne*[13] or *ne...not*.[14] A pair of discourse-new propositions adjacent to each other in a text are not more likely to have the same negative marker than two discourse-new propositions selected at random from the dataset. Discourse-old propositions behave similarly: 67.5% (n=176/260) of discourse-old propositions are predicted to have the

[13] $\chi^2$(1df)=0.239, p=.625.
[14] $\chi^2$(1df)=2.99, p=.084.

Table 6.10. *Comparison of predicted and observed frequencies of priming in the period 1150–1250*

| | Discourse-new | | | Discourse-old | | |
|---|---|---|---|---|---|---|
| | *ne* | *ne...not* | Total | *ne* | *ne...not* | Total |
| Overall | 86.4% (n=352/407) | 13.5% (n=55/407) | 407 | 20.4% (n=53/260) | 79.6% (n=207/260) | 260 |
| Predicted priming (overall frequency[2]) | 74.7% (n=304/407) | 1.8% (n=7/407) | **76.5%** (n=311/407) | 4.1% (n=11/260) | 63.4% (165/260) | **67.5%** (n=176/260) |
| Observed priming | 76.1% (n=310/407) | 3.7% (n=15/407) | **79.8%** (n=325/407) | 3.4% (n=9/260) | 61.5% (n=160/260) | **65.0%** (n=169/260) |

Table 6.11. *Comparison of predicted and observed frequencies of priming in two texts from the period 1250–1350.*

| Discourse-new | The Ayenbite of Inwit | The Earliest Prose Psalter |
|---|---|---|
| Predicted same as preceding clause | 55.5% (n=102/183) | 58.6% (n=92/157) |
| Observed same as preceding clause | 53.0% (n=97/183) | 63/0% (n=99/157) |
| $\chi^2$ | 0.275 | 0.66 |
| p= | .60 | .42 |
| **Discourse-old** | **The Ayenbite of Inwit** | **The Earliest Prose Psalter** |
| Predicted same as preceding clause | 72.1% (n=94/130) | 78.1% (n=170/218) |
| Observed same as preceding clause | 74.6% (n=97/130) | 78.9% (n=172/218) |
| $\chi^2$ | 0.18 | 0.05 |
| p= | .67 | .82 |

same NM as any other randomly selected discourse-old proposition. In fact, we find that 9 clauses with *ne* and 160 clauses with *ne...not* have the same NM as the preceding clause, giving a total of 65.0% (n=169/260). The observed and predicted totals are not significantly different ($\chi^2$(1df)=0.422, p=.516), neither are the predicted and observed frequencies of *ne*[15] or *ne...not.*[16] A pair of discourse-old propositions adjacent to each other in a text are no more likely to have the same negative marker than two discourse-old propositions selected from the dataset at random. Therefore there is no independent priming effect. The effect is observed by Ecay and Tamminga derived from the independent distributions of forms when we take the different functional contexts into account.

In 1250–1350, the period Ecay & Tamminga focus on, they find the form of negation is the same in the prime and the target 63.8% of the time. Does this follow from the overall distributions of *ne*, *ne...not* and *not* in the dataset without appealing to a priming effect? There are only 3 texts dating from 1250–1350 in the PPCME2 corpus, and they are quite different, indicating we should separate them in the analysis. Two of the texts, the *Ayenbite of Inwit* and the *Earliest Prose Psalter* provide sufficient tokens to be analysed individually.

The predicted and observed frequencies of priming in each discourse context and text are compared in Table 6.11, along with $\chi^2$-tests estimating the significance of differences between the predicted and observed figures in

15 $\chi^2$(1df)=0.208, p=.648.
16 $\chi^2$(1df)=0.205, p=.651.

each context. Once we take into account the distinction between discourse-old and discourse-new propositions and differences between the individual texts, there is no evidence for a priming effect in either text. Whether or not two clauses are adjacent makes no difference to the likelihood that they will share the same NM. Two clauses are just as likely to share the same NM whether they are adjacent or selected at random from each text, provided they share the same discourse-function.

The priming effect that Ecay & Tamminga find is therefore a consequence of discourse function, and arises because they do not distinguish activated (discourse-old) and non-activated (discourse-new) propositions within their model. This provides a further demonstration that writers treat discourse-old and discourse-new propositions as distinct contexts for negation strategies. Once discourse-functions are taken into account, we can derive the priming effect in a model where negative markers are independently distributed rather than primed. The 'priming effect' is reducible to the difference between discourse-old and discourse-new propositions that we already have in the model. It is not the form of the negative marker in the prime that is the important determinant of the negative marker in the target, but whether the target is a discourse-new or discourse-old proposition.

## 6.6 Implications: Functional Constraints on Present-day English not-negation

The pragmatic constraints on ME *not* persist into PDE in contexts where *not* remains optional. There are two ways to negate clauses with post-verbal indefinite elements in Present-day English: no-negation, as in (137a); and not-negation, as in (137b).

(137) a. I said nothing (no-negation)
b. I didn't say anything (not-negation)

*not* is not grammaticalised in these clauses, but remains variable. If the mechanism that introduces *not* into these clauses is the same as in sentential negation contexts (the Jespersen Cycle), but simply does not go to completion, then not-negation should be subject to the same pragmatic constraints as pre-grammaticalised *not* in other contexts. Furthermore, if pragmatic constraints on not-negation persist even into Present-day English, they provide further evidence that *not* only becomes pragmatically unmarked when it becomes the grammatical default – that is, when its use is categorical.

Like Middle English *not*, Early Modern English not-negation tends to occur in contexts of pragmatic activation like those in (138)–(141).

(138) Repetition of antecedent proposition:
L. C. J. No, thou saidst it was so dark <u>thou could'st not see any thing</u>

'L.C.J.: No, you said it was so dark you couldn't see anything'
(17th century, LISLE-E3-P2,4.114.17)

(139) Denial of an antecedent proposition:
Why Sir Tho. Peyton should tell anie bodie that I refused a liveing of a $=li= by the yeare I knowe not ; for I have not refused anie 'Why Sir Thomas Peyton should tell anybody that I refused a living of a 5l per year I don't know; because I have not refused anything'
(17th century, HOXINDEN-1660-E3-H,276.88–89)

(140) Cancellation of an inference:
L.C.J. Did he pull down the Hay or you ? (LISLE-E3-H,IV,114C2.104)
Dunne. I did not pull down any Hay at all. (seventeenth century, LISLE-E3-H,IV,114C2.105)

(141) Statement of an inference:
as for Mis Kettey : I prays the lord we have noe neede of Mr. Annett : for shee never had any spice of an ill fitt since shee came to mee last ; and also she mends dayly in her stomake for her victuals. As for her worke , she shall doe as little as may bee of one thinge : for her travilling with mee to barham : I would not venture her for a weeke : neither doe I perceive that shee desires itt : and for my selfe I shall hardly cume at all to broome : or if I doe I shall stay but one day and so hombe [home] agayne : and be Asured that mis Kettey shall not loose any tyme, only what is convenient for helthfull Recreations: (seventeenth century, JACKSON-E3-P1,206.18–27)

In (138) the underlined proposition is reported speech, repeated from the preceding dialogue, while in (139) the proposition 'I have not received anie' contradicts the preceding proposition. The proposition 'I did not pull down any hay at all' in (140) cancels the presupposition introduced by the preceding question that someone did in fact pull down some hay. Finally, the proposition 'Miss Kettey shall not lose any time' in (141) can be inferred from the fact that the writer and Miss Kettey will only spend one day away on their travels.

The same pragmatic distinction is also evident in late twentieth-century spoken data. In spoken data from the BNC, when the proposition being negated already has a truth value (true or false) assigned to it either explicitly in the preceding discourse, or when its truth value can be inferred from the preceding discourse, not-negation tends to appear – so not-negation is in a sense emphatic, reinforcing or contradicting a proposition whose truth-value is already present in the discourse context, as in the PDE example (142) from the BNC.

(142) SP:PS087 What do you want to watch? ...
SP:PS089 [something] I haven't watched before.
(presupposing that there *is* something new)
SP:PS087 There isn't anything new, no.

On the other hand, when the truth value of the negative proposition cannot be recovered from the discourse, but is assigned by the negative proposition itself for the first time, no-negation tends to occur, as in the PDE example (143) from the BNC.

(143) SP:PS1HJ See anything?
(neutral with respect to to the truth-value of the question)
SP:PS1HH: No, I looked for rabbits and squirrels. I could see nothing.

Table 6.12 presents the distribution of not-negation according to discourse-function in two datasets, subdividing the data into the same five functions that we identified in ME. The first dataset comprises PPCEME data from the period 1640–1710. These data all represent the written register, though there are examples of reported speech in the corpus for example trial records, and also constructed dialogues in dramatic texts. The second dataset comprises spoken conversation from the British National Corpus dating from c.1990–1993.[17] However, in order to restrict the BNC dataset to manageable proportions, only post-verbal indefinite *-thing* pronouns are analysed, as in (142) and (143). The PPCEME data are not restricted in this way, and include all post-verbal arguments with indefinite *any* or negative *no* as well as indefinite and negative pronouns.

The distributional analysis (Table 6.12) shows a clear distinction between discourse-given and discourse-new contexts in both datasets. The table excludes main verbs HAVE and BE, as not-negation is very infrequent with these verbs (see Tottie (1991b) and the discussion in Section 9.4.2). In the BNC spoken conversation subsample, we see that the frequency of not-negation (142) is almost at ceiling in three of the four discourse old contexts, whereas it is much lower in the discourse-new propositions.[18]

Comparison of these data with those in Section 6.3.3 suggest historical persistence of functional constraints on *not*, from Middle English through to Present-day English not-negation. The distribution of Early Modern English not-negation is is reminiscent of the distribution of *ne...not* in Early Middle English. In Table 6.13 a series of mixed-effects regression models compare the probabilistic effect of the discourse-givenness/newness constraints on not-negation and no-negation in the seventeenth-century data and on the competition between *ne* and *ne...not* in the early Middle English PPCME2

[17] In both datasets, both examples of negative doubling with *not* and examples with *not* + post-verbal NPI are counted as not-negation.

[18] These BNC data contain both instances of not-negation in which the indefinite element is a negative polarity item *anything* and in which it is a concordant negative item *nothing*. A logistic regression model comparing the distribution of these two types of not-negation across discourse functions and verb types reports no statistically significant ($p \leq .05$) differences in their distributions.

Table 6.12. *Distribution of no-negation and not-negation in PPCEME data for the period 1640–1710 and BNC spoken conversation subsample,* HAVE *and* BE *excluded.*

| Function | PPCEME data no-negation | not-negation | Total | BNC conversation no-neg | not-neg | Total |
|---|---|---|---|---|---|---|
| Denial of antecedent | 66.7% (n=4) | 33.3% (n=2) | 6 | 4.7% (n=3) | 96.3% (n=61) | 64 |
| Repetition of antecedent | 55.6% (n=5) | 44.4% (n=4) | 9 | 14.1%[19] (n=10) | 85.9% (n=61) | 71 |
| Assertion of inference | 61.1% (n=11) | 38.9% (n=7) | 18 | 3.8% (n=3) | 96.2% (n=76) | 79 |
| Cancellation of inference | 48.6% (n=17) | 51.4% (n=18) | 35 | 0.8% (n=1) | 99.2% (n=128) | 129 |
| Sub-total: discourse-given | 54.4% (n=37) | 45.6% (n=31) | 68 | 5.0% (n=17) | 95.0% (n=326) | 343 |
| Discourse-new | 95.0% (n=134) | 5.0% (n=7) | 141 | 35.1% (n=68) | 64.9% (n=126) | 194 |
| TOTAL | 81.8% (n=171) | 18.2% (n=38) | 209 | 15.8% (n=85) | 84.2% (n=452) | 537 |

Table 6.13. *Comparison of the constraints on Early Modern English no-negation (data from Table 6.12) and not-negation with those on Middle English ne and ne...not (data from Table 6.4).*

| | not-negation vs. no-negation 1640–1710 | | *ne...not* vs. *ne* 1150–1250 | |
|---|---|---|---|---|
| | Effect | p | Effect | p |
| Repetition of proposition | 2.93 | .001 | 2.62 | .001 |
| Statement of inference | 2.22 | .001 | 3.38 | .001 |
| Cancellation of inference | 3.15 | .001 | 3.43 | .001 |
| Denies a proposition | – | – | 2.78 | .001 |

data. The regression analysis includes discourse-function as a fixed effect (predictor) and corpus text as a random effect to incorporate any differences between individual texts into the model.[20]

The effect sizes for the four discourse-old proposition types are estimated via treatment coding. The effect sizes estimate, in logits, the extent to which the distribution of *not* or *ne...not* differs between these contexts and

[19] The higher frequency of no-negation in this context might be the result of a priming effect. If the proposition that is repeated as no-negation, speakers may tend to use no-negation when repeating it more often than in the other discourse-given contexts.

[20] As discussed in Chapter 2, this is important because the the dataset is unbalanced. The corpus is not homogenous, but comprises different texts, each of which contributes a different proportion of the overall dataset.

Table 6.14. *Comparison of the discourse-new/discourse-old contrast on the distribution of not in three datasets from different periods.*

| Dataset | Discourse-old | p |
|---|---|---|
| PPCME2 *ne..not* vs *ne*, written, 1150–1250 | 3.01 | .001 |
| PPCEME not-negation/no-negation, written, 1640–1710 | 2.79 | .001 |
| BNC, not-negation/no-negation, conversation, 1985–1993 | 2.41 | .001 |

discourse-new propositions (the reference level). Positive effect sizes indicate propositions which favour *not* or *ne...not* over discourse-new propositions. Gaps in the table represent contexts with too few (n<4) tokens for reliable regression analysis. The high p-values here indicate that differences between discourse-given and discourse-new contexts are highly significant in both datasets.

Although the overall frequency of *not* is much lower[21] in the seventeenth-century data with indefinites (18.2%, n=38/209) than *ne...not* is in the early ME sentential negation data (39.2%, n=264/674); in both datasets, the direction and strength of discourse constraints are similar. In each dataset, there are no statistically significant differences between each of the discourse-old contexts. This indicates that we can reduce discourse constraints to a simple two way contrast between discourse-old propositions (which all pattern alike) and discourse-new propositions. This allows us to compare the discourse-old/new contrast in ME written data, seventeenth-century written data from the PPCEME, and contemporary spoken BNC data. In the BNC sample, there are too few instances of no-negation in each discourse-old context to perform regression analysis without aggregating the four discourse-old contexts into a single group.

Irrespective of the different overall frequencies of not-negation in spoken versus written registers, no-negation and not-negation remain subject to the same discourse constraints as *ne* and *(ne)...not* are in Early Middle English sentential negation contexts. The pragmatic functions associated with Early Modern English and Present-day English not-negation reflect the pragmatic functions of *not* at the earliest stages of the Jespersen Cycle, when *not* is optional, and before *ne* is lost. They are clear evidence against the pragmatic unmarking of *not* in ME at stage three of the Jespersen Cycle. Pragmatic unmarking of *not* happens only when its use becomes categorical. In clauses with indefinites where not-negation remains variable, we see historical persistence of the earlier pragmatic constraints on its use over a period of 800 or more years.

[21] Significantly lower: $\chi^2(1df) = 31.23$, $p<.001$.

This persistence is most easily explained if the spread of not-negation is a consequence of the more general introduction of *not* as part of the ME Jespersen Cycle, rather than a separate development, but this raises the question of why the spread of not-negation does not go to completion. Clauses with indefinites are the only context in which *not* remains optional after $ne_1$ is lost. I return to this question in Chapter 9.

## 6.7 Conclusions

Middle English *not* is subject to discourse-functional constraints on its distribution based on pragmatic activation, irrespective of whether it occurs alone or whether it co-occurs with *ne*. In propositions that are new to the discourse – those which are not identifiable or inferable from the earlier discourse – *ne* is much more likely to appear than *(ne)...not*. Conversely, in propositions that are discourse-old – that is, already activated and present to the reader's attention through being previously mentioned, or inferred from the earlier discourse – *(ne)...not* is much more likely to mark sentential negation. This functional differentiation of successive stages in the ME Jespersen Cycle helps to explain the rather extended period of variation between *ne* and *ne...not/not* we find in Middle English from the twelfth to fourteenth centuries. The effects of pragmatic activation on the distribution of ME *not* parallel those described for bipartite negative strategies in Romance languages such as Catalan *no...pas* (Schwenter, 2006), French *ne...pas* (Hansen, 2009) and Italian *non...mica* (Hansen and Visconti, 2009). The ME data show that the pragmatic constraints extend to stage three negation strategies too, such as *not* while these stage three strategies compete with stage one strategies like *ne*.

Hansen (2009) and Hansen and Visconti (2009) hypothesise that French bipartite *ne...pas* becomes pragmatically unmarked over time. A distributional analysis of competition between ME *ne* and *ne...not* lends support to the idea that a similar process of pragmatic unmarking happens in Middle English. The contrast between discourse-old and discourse-new propositions appears to decrease as *ne...not* becomes more frequent. However, the frequency data do not require an analysis in terms of pragmatic unmarking, and in fact do not indicate whether pragmatic unmarking is taking place or not.

Within a regression analysis, pragmatic unmarking should manifest itself as a weakening of the pragmatic constraints on competition between *ne* and *(ne)...not* over time. A mixed-effects regression model does not demonstrate pragmatic unmarking at the constraint level. When the strength of each pragmatic constraint on variation between *ne* and *ne...not* is estimated through regression analyses of data from two successive time periods, we see that the effect of the discourse-old/discourse-new distinction is consistent across these periods, despite the increasing overall frequency of *(ne)...not*. The increasing overall frequency of *(ne)...not* appears independent of the

Table 6.15. *Interaction of formal, semantic and pragmatic changes in the Jespersen Cycle.*

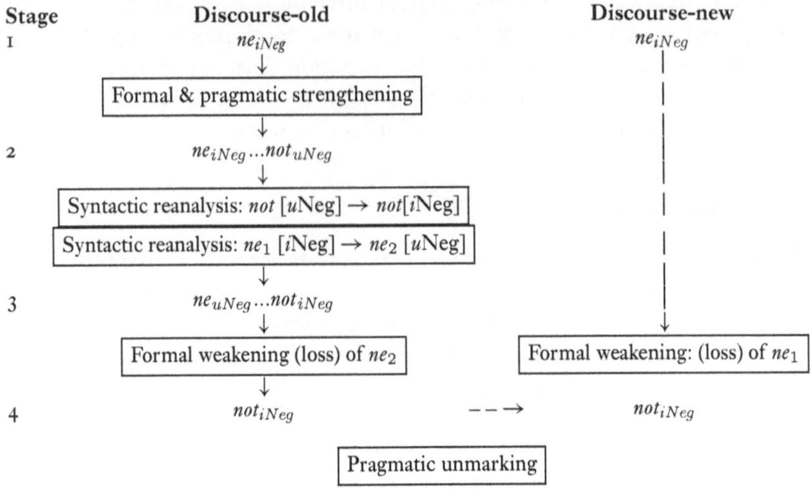


functional constraints on its use. Hence the apparent loss of the contrast between discourse-old and discourse-new contexts in the frequency data simply results from competition between $ne_1$ and *not* progressing along a logistic curve. The slope of the curve – the rate of change – is the same in both discourse contexts. The constraining effect of pragmatic activation remains constant in both periods, irrespective of the overall increase in use of *(ne)...not*. Therefore, accommodating pragmatic activation within a logistic model of the Jespersen Cycle is straightforward – pragmatic activation simply supplies constraints on the distributions of $ne_1$ and *(ne)...not* that remain constant over time. We cannot appeal to weakening of these particular pragmatic activation constraints to explain the spread of *ne...not* and *not* in the thirteenth and fourteenth centuries. Hence there is no evidence to discern pragmatic unmarking at the constraint level.

Instead, *not* only becomes grammaticalised as the pragmatically unmarked grammatical default negative marker in the fifteenth century, after *ne* is lost. This creates pressure to generalise *not* into contexts where *ne* formerly marked negation, simply in order that negation be marked. The statistical data presented in this chapter are evidence for a model of the Jespersen Cycle in which pragmatic and syntactic changes interact as illustrated in Table 6.15. While changes in discourse-new and discourse-old contexts may follow either of the two trajectories outlined in Table 6.15, in probabilistic terms, one trajectory is highly favoured in discourse-old propositions while the other is highly favoured in discourse-new propositions.

The patterning of discourse-functional constraints at each stage of the Jespersen Cycle provides further evidence for the conclusion reached in

Chapters 2–5 – that the distributions of *ne*$_1$ (Jespersen Cycle stages one/two) and *ne*$_2$ (stage three) are independent of each other. While competition involving *ne*$_1$ is highly constrained by discourse-function, the effect of discourse function on *ne*$_2$ patterns differently, and the effect is much weaker. The result of competition between *ne*$_1$ and *ne*$_2$ is that both *ne* and *not* appear to undergo pragmatic unmarking in the fourteenth century when *ne*$_1$ disappears. However, functional change simply follows as a consequence of competition within the syntactic model I proposed in Chapter 5. A more functionally articulated model of the kind schematised in Table 6.15 also subsumes the priming effects noted by Ecay and Tamminga (2016) , rendering independent priming constraints unnecessary.

There are two key points in Table 6.15. First, that discourse-old and discourse-new contexts favour different negation strategies until *ne*$_1$ is lost in the late fourteenth century. There is no pragmatic unmarking of the *ne...not* construction. Second, the syntactic reanalysis of *not* as a negative marker occurs in pragmatically marked contexts and has no effect on the pragmatic constraints on *(ne)...not*. The Middle English data therefore provide evidence, contrary to accounts such as Detges and Waltereit (2002), Hansen (2009), Hansen and Visconti (2009), that syntactic reanalysis is independent of – and can precede – pragmatic unmarking. The syntactic reanalysis of *not* from minimiser to negative marker in contexts of pragmatic activation, sets up a situation of competition between syntactically and pragmatically distinct options which plays out over a long period of time in a way that is consistent with the Constant Rate Effect.

# 7 Negative Concord in Early English

## 7.1 Introduction

The following three chapters examine the distribution of early English negative concord in diachronic corpus data. This chapter provides an overview of early English negative concord. Negative concord is not a homogenous phenomenon in early English – there is both diachronic and diatopic variation in the types of negative concord employed. Giannakidou (2000, 2006) identifies two parameters of negative words in negative concord languages: their negativity – whether or not negative words contribute negative force at LF, and their quantificational force – whether they are universal quantifiers, existential quantifiers or indefinites. I argue that, in order to derive the patterns of negative concord we find in early English and the changes in their availability over time, negative arguments and adverbials vary in precisely these two properties.

Van der Wouden (1994) distinguishes two types of negative concord: negative doubling involves concord between a negative marker and one or more negative words, as in (144); negative spread involves concord between two or more negative words but no negative marker, as in (145).

(144) Ic ne funde nanne gylt on him
I NEG found no sin in him
'I found no sin in him' (cowsgosp,Jn_[WSCp]:19.6.7261)

(145) no man seyd no-thyng a-gens hem
no man said nothing against him
'no man said anything against him' (CMKEMPE,33.730)

Chapters 8 and 9 investigate the relationship between negative doubling and the model of the Jespersen Cycle I proposed in Chapters 2–6. The relationship between negative doubling and the Jespersen Cycle is not a straightforward one. In Old English and Early Middle English, almost all clauses with negative words exhibit negative doubling with *ne* – the pattern illustrated by (144). However, at stages three and four of the Jespersen Cycle, the negative marker *not* does not take over from *ne* in negative doubling contexts in the way that it does in sentential negation contexts. Instead, negative doubling begins to be lost. Consequently, the use of negative doubling during ME becomes variable. Chapter 9 argues that variability between PDE no-negation (146a) and not-negation (146b) originates in the interplay between the loss of negative concord and the Jespersen Cycle during ME and persists unchanged into Present-day English.

(146) a. No-negation: I saw nothing
 b. Not-negation: I didn't see anything

I demonstrate that although their effects manifest themselves in different ways in sentential negation and negative doubling, the same processes that underlie the Jespersen Cycle are responsible for changes in negative doubling during ME; and, by extension, for variation between no-negation and not-negation in PDE. Evidence from change in progress in diachronic corpus data indicates that the ME Jespersen Cycle and the ME loss of negative concord are both reflexes of the same two changes – competition between *ne*1 and *ne*2, and the subsequent loss of *ne*2.

## 7.2 Negative Doubling in Early English

Negative doubling is typical of early English from the eighth century until the 14th century. Clauses with negative words such as a quantifiers, pronouns or adverbs usually also employ the negative marker *ne*, as in (147).

(147) Ne mæg þonne nan man nahwar beon behydd
 NEG can then no man nowhere be hidden
 'Then no man can be hidden anywhere'
 (coaelhom,ÆHom_11:391.1688)

Table 7.1 reports the frequency of negative doubling in clauses with negative arguments and adverbials like (147). These figures include negative

Table 7.1. *The frequency of negative doubling with negative arguments and adverbials in three diachronic corpora.*

| Period | negative doubling | Total | % negative doubling |
|---|---|---|---|
| YCOE: | | | |
| 850–950 | 1104 | 1320 | 83.6% |
| 950–1050 | 2908 | 3099 | 96.4% |
| 1050–1150 | 1102 | 1230 | 89.6% |
| PPCME2: | | | |
| 1150–1250 | 757 | 806 | 93.9% |
| 1250–1350 | 183 | 205 | 89.3% |
| 1350–1420 | 167 | 1296 | 12.9% |
| 1420–1500 | 36 | 1392 | 2.6% |
| PPCEME: | | | |
| 1500–1570 | 24 | 1433 | 1.7% |
| 1570–1640 | 5 | 1620 | 0.3% |

doubling with *ne*, *ne...not* and *not*. The figures include main clauses, second conjuncts and subordinate clauses in three corpora. The YCOE (Taylor et al., 2002) provides the data for 850–1150. The PPCME2 (Kroch and Taylor, 2000) provides the data for 1150–1500, and the PPCEME (Kroch et al., 2004) the data for 1500–1710.

There is some variability in the occurrence of negative doubling throughout the Old English and Middle English periods. However, a steep decline in negative concord occurs during the fourteenth and fifteenth centuries. What accounts for the variability in negative doubling prior to the point where negative doubling begins to be lost in the mid fourteenth century, and how can we account for the loss of negative doubling from the mid fourteenth century onwards?

## 7.3 Strict and Non-strict Negative Doubling

Much of the variation in the frequencies of negative doubling in the Old English (850–1150) and early Middle English (1150–1350) periods may be dialectal. Ingham (2006) describes a pattern of variation between strict and non-strict negative concord dialects in Old and Middle English. He argues that West Saxon Old English exhibits symmetrical negative doubling, such that a negative quantifier occurs in negative doubling with *ne* irrespective of the relative positions of the two elements. In (148) the negative quantifier precedes *ne* and in (149) the negative quantifier follows *ne*.

(148) & nan mann ne mihte þæs weges faran.
and no man NEG can this-GEN way-GEN travel
'and no man can travel this way'
(coaelhom,ÆHom_18:225.2608)

(149) And hi þa ne mihton nan þing him geandwyrdan
and they then NEG could no thing him answer
'and then they could not give him any answer'
(YCOE, coaelhom,ÆHom_2:274.385)

This looks like strict negative doubling. However, Ingham argues that non-West-Saxon texts exhibit a different pattern – negative doubling occurs when a negative marker precedes the negative quantifier, but negative doubling is absent when the negative quantifier precedes the negative marker, as in (150). This looks like non-strict negative doubling (cf. Giannakidou (2006, 23ff)).

(150) a. þæt he nane þinga him andwerdan wolde
that he no things him-DAT answer would
'that he would in no wise answer him'
(Wærferth 122, 16, Ingham (2006, 249, ex.13))

b. & he næfre ma wonað
and it [sc.: the sun] never more sets
'and it will never set'
(Martyrology 42, 25, Ingham (2006, 249, ex.16))

### *7.3.1 Strict and Non-strict Negative Doubling Dialects*

Ingham describes this variation in terms of the relative positions of negative quantifiers and negative markers. However, the distinction between strict and non-strict negative concord is a structural one, based on scope and described in terms of the structural relation of C-command, which must hold at S-structure (Giannakidou, 2006, 36). In non-strict negative doubling the negative word must remain within the scope of (C-commanded by) the negative marker at S-structure. In Old English, *ne* is proclitic on the finite verb and therefore appears in a variety of different surface positions. Given that *ne* is proclitic on the finite verb in Old English, and the landing site of the finite verb may be head-initial or head-final (Pintzuk, 1999), a pre-verbal negative quantifier may in fact remain within the scope of *ne* when *ne* is a clitic on a finite verb located in a head-final TP, provided the negative quantifier is in the complement of $T^0$, as in examples like (151).

(151) ic nanne intingan findan ne mæg on þisum men
I no case find NEG can against these men
'I can find no case against these men'
(cowsgosp,Lk_[WSCp]:23.13.5567)

This may account for the fairly high incidence of negative doubling Ingham observes with pre-verbal negatives even in non-West-Saxon texts – Ingham (2006, 248, Table 1) gives a figure of 38.5% (n = 47/122). Although negative constituents occur pre-verbally in examples like (151), they may in fact remain in the scope of a head-final TP/AgrP, and therefore comprise contexts for non-strict rather than strict negative doubling.

Examining patterns of Old English negative doubling in head-initial clauses ensures that negative quantifiers preceding finite verbs must be outside the scope of *ne*. These are clauses in which there is a finite verb preceding a non-finite verb, like (152).

(152) a. nan man ne sceal sceattas niman for Godes cyrcan
no man NEG ought property sieze for Gods-GEN church
'no man ought seize property for God's church'
(cocathom2,ÆCHom_II,_45:344.293.7705)

b. We witon ðæt nan man mæg nawyht goodes wyrcan, buton
We know that no man can nothing good-GEN do, but
hym God myd wyrce;
him God with work-SUBJUNCTIVE
'We know that no man may do anything of good unless he work with God'
(cosolilo,Solil_1:30.22.397)

Given the very low rate of postposing for non-finite verbs and VPs (Pintzuk, 1999, 114ff), these word orders are typically derived by movement of the finite verb to a head initial $T^0$ or $Agr^0$. Thus any negative argument or adverbial preceding the finite verb has also undergone movement to a position not

C-commanded by the finite verb. Restricting the dataset in this way provides a much smaller number of examples than in Ingham's dataset, but produces broadly similar results. Negative elements preceding a finite verb in $T^0/Agr^0$ occur in negative doubling with *ne* in most texts. The exceptions – negative quantifiers preceding a head initial $T^0/Agr^0$ without negative doubling are shown in Table 7.2. These only occur in the Old English *Bede* (153a), the *Benedictine Rule* (153b), the life of St Chad (153c), the Old English *Cura Pastoralis* or Gregory's Pastoral Care (153d)[1], the Old English translation of Gregory's Dialogues (ms. C) (153e), the Laws of Ine (153f), the Life of St. Margaret (ms. C) (153g), the Old English Martyrology (153h) and King Ælfred's version of Augustine's Soliloquies (153i).

(153) a. & nænig mon dorste for hine ne sealmas ne mæssan
and no man dare for him neither psalms nor masses
singan
sing
'and no man dare sing either psalms or masses for him'
(cobede,Bede_5:15.444.10.4462)

b. and nænig sy beladod fram þære kycenan þenunge
and none be-SUBJ excused from the-GEN kitchen-GEN service
'and none be excused from kitchen duties'
(cobenrul,BenR:35.58.13.718)

c. Þeah we nenge þinga magen becuman to eallum þam
yet we no thing-DAT may come to all the
megenum his weorca
strength his works-GEN
'Yet in no way may we meet with all the power of his works'
(cochad,LS_3_[Chad]:4.4)

d. To ðæm gebanne ðæs tohopan nan monn mæg cuman,
To the summons that hope-GEN no man may come,
butan he ðider ierne mid anmodnesse wið his niehstan.
unless he thither run with concord with his neighbour
'To the summons of that hope no man may come, unless he run thither with concord towards his neighbour'
(cocura,CP:46.345.19.2325)

e. swa þæt nanra þinga mid ænigre efestinge mannes hi
so that no thing with any haste man-GEN they
mihton beon undon & unwriðene.
could be undone and unbound
'so that in no way could they be undone by any man's haste'
(cogregdC,GD_2_[C]:31.164.8.1973)

[1] There are two manuscript variants of this example in Sweet's edition (Sweet, 1871): The example as found in the corpus comes from the Hatton ms. and lacks negative doubling with *ne*, but in the Cotton ms., the same example exhibits negative doubling with *ne*.

f. þætte nænig ealdormonna ne us undergeðeodedra æfter þam
that none aldermen-GEN nor us subjects after that
wære awendende ðas ure domas.
were-SUBJ spared this our judgement
'that none of the aldermen nor of us subjects were to be spared our judgement after that'
(colawine,LawIne:0.1.2)

g. þæt innan heora husum nan unhal cild sy geboren
that in their house no sick child be-SUBJ born
'that in their house no sick child be born'
(comargaC,LS_14_[MargaretCCCC_303]:19.13.322)

h. þæt nænig wer hi scolde geseon butan him anum,
that no man they should see but him alone
'that they should see no man but him alone'
(comart3,Mart_5_[Kotzor]:Jy19,A.5.1191)

i. We witon ðæt nan man mæg nawyht goodes wyrcan, buton
We know that no man can nothing good-GEN do, but
hym God myd wyrce;
him God with work-SUBJ
'We know that no man may do anything of good unless he work with God'
(cosolilo,Solil_1:30.22.397)

Although the numbers here are very small, they offer some support for the pattern identified by Ingham (2006) – negative doubling shows a tendency towards being asymmetric in these texts, in contrast to the more general preference for symmetric negative concord in Old English. Ingham concludes that we have two distinct negative concord dialects in Old English: a predominantly West Saxon dialect with symmetric negative concord, and a predominantly Mercian or Anglian dialect with asymmetric negative concord.[2] The strict–non-strict distinction accounts for nearly all the variation in negative doubling we find in YCOE prose texts. Negative doubling is categorical with post-verbal negative words in all but one text[3], but variable with pre-verbal negative words depending on text and dialect.

If, as Ingham argues on the basis of thirteenth-century verse texts, this variation between symmetric and asymmetric negative concord dialects persists into Middle English, it needs to be taken into consideration as we account for the loss of negative doubling. The frequencies of negative doubling with pre-verbal and post-verbal negatives need to be considered

[2] Of the texts in Table 7.2, *Bede, Life of St. Chad* and *Gregory's Dialogues* are identified as possibly Anglian/Mercian in origin by the compilers of the Helsinki and YCOE corpora. Rauer (2013) suggests an Anglian origin for the OE *Martyrology*. However, we should note that the OE *Cura Pastoralis* is thought to be an exponent of West Saxon (Fulk, 2012, fn.1).

[3] *Augustine's Soliloquies* provides 2 exceptions.

Table 7.2. *The frequency of negative doubling with negative quantifiers in texts where it is not categorical. All other YCOE texts exhibit 100% negative doubling in both contexts.*

| | not C-commanded by Vf | | | C-commanded by Vf | | |
|---|---|---|---|---|---|---|
| Text | *ne* | Total | % *ne* | *ne* | Total | % *ne* |
| Bede | 2 | 7 | 28.6% | 3 | 3 | 100.0% |
| Benedictine Rule | 1 | 3 | 33.3% | 11 | 11 | 100.0% |
| Life of St. Chad | 0 | 1 | 0.0% | 0 | 0 | – |
| OE Cura Pastoralis | 7 | 8 | 87.5% | 13 | 13 | 100.0% |
| Gregory's Dialogues (C) | 10 | 12 | 83.3% | 29 | 29 | 100.0% |
| Laws of Ine | 0 | 1 | 0.0% | 0 | 0 | – |
| Life of St. Margaret | 0 | 1 | 0.0% | 1 | 1 | 100.0% |
| OE Martyrology (3) | 0 | 2 | 0.0% | 5 | 5 | 100.0% |
| Augustine Soliloquies | 4 | 5 | 80.0% | 12 | 14 | 85.7% |

separately across time in order to distinguish potential dialectal variation from diachronic change.

The patterns in the PPCME2 data are much less clear than in Ingham's (2006) verse texts. Table 7.3 reports the frequency of negative doubling with pre- and post-verbal negative quantifiers in periods from 1150–1420.[4] In the period 1150–1250, there is very little evidence for non-strict negative concord. The majority of texts in this sub-period of the PPCME2 are localised to the West Midlands and South West. These western texts typically maintain strict negative concord (the *Katherine group, Ancrene Riwle, Vices and Virtues*). Those texts showing variation in negative doubling with pre-verbal negatives show the same pattern with post-verbal negatives, suggesting variability in the use of strict negative doubling rather than strict/non-strict asymmetry (for example in the *Peterborough Chronicle* and *Trinity Homilies*).

Moving on to the period 1250–1350, we find some evidence for asymmetric negative concord particularly in the *Early Prose Psalter*, which is localised to the East Midlands dialect. However, the works of Rolle (localised to Yorkshire) suggest that negative doubling is lost both pre- and post-verbally in texts of more northern origin.

In the period 1350–1420, the situation is variable and less clear. The loss of negative doubling appears to occur rather abruptly in the fourteenth century. Negative doubling is only maintained in a few texts, principally texts by Chaucer and Mandeville, which Jack (1978a) argues represent a particular South East Midlands or London dialect. We find symmetric (strict) negative doubling in Chaucer's Parson's Tale (London), and Boethius translation (London)[5], in Mandeville's Travels (South East) and Laymon's

[4] As the frequency of head-final TP/AgrP is quite low (around 5 per cent of clauses) in Middle English (Kroch and Taylor, 2000, 143), the figures include all clauses, not just those with a finite verb preceding a non-finite verb (in contrast to the Old English figures in Table 7.2).

[5] The frequency of negative doubling seems much higher in Chaucer's Boethius than the other texts by Chaucer. One possibility is that this is a translation effect, and that patterns of negative doubling in this text reflect the French sources Chaucer used.

Table 7.3. *Patterns of negative doubling (ND) involving ne, ne...not or not with pre-verbal and post-verbal negative elements in PPCME2 texts.*

| | | Pre-verbal negative | | | Post-verbal negative | | |
|---|---|---|---|---|---|---|---|
| Period | Text | ND | Total | % ND | ND | Total | % ND |
| 1150–1250 | Ancrene Riwle | 20 | 20 | 100.0% | 28 | 30 | 93.3% |
| | Hali Meidhad | 5 | 5 | 100.0% | 16 | 17 | 94.1% |
| | Katherine Group[6] | 10 | 10 | 100.0% | 42 | 43 | 97.7% |
| | Kentish Homilies | 2 | 2 | 100.0% | 9 | 9 | 100.0% |
| | Lambeth Homilies | 9 | 10 | 90.0% | 26 | 27 | 96.3% |
| | Peterborough Chronicle | 10 | 12 | 83.3% | 4 | 8 | 50.0% |
| | Trinity Homilies | 18 | 21 | 85.7% | 33 | 35 | 94.3% |
| | Vices and Virtues | 42 | 42 | 100.0% | 39 | 39 | 100.0% |
| | Total | 116 | 122 | 95.1% | 197 | 208 | 94.7% |
| 1250–1350 | Ayenbite of Inwit | 56 | 58 | 96.6% | 62 | 62 | 100.0% |
| | Early Prose Psalter | 0 | 6 | 0.0% | 9 | 14 | 64.3% |
| | Kentish Sermons | 5 | 5 | 100.0% | 3 | 3 | 100.0% |
| | Rolle Epistles | 0 | 30 | 0.0% | 1 | 49 | 2.0% |
| | Rolle Treatises | 0 | 9 | 0.0% | 1 | 38 | 2.6% |
| | Total | 61 | 108 | 56.5% | 76 | 166 | 45.8% |
| 1350–1420 | Ældred of Rievaulx (Vernon ms.) | 0 | 23 | 0.0% | 1 | 28 | 3.6% |
| | Astrolabe | 0 | 1 | 0.0% | 0 | 6 | 0.0% |
| | Rule of St Benet | 0 | 36 | 0.0% | 8 | 57 | 14.0% |
| | Chaucer Boethius | 4 | 5 | 80.0% | 15 | 23 | 65.2% |
| | Brut | 2 | 46 | 4.3% | 0 | 52 | 0.0% |
| | Cloud of unknowing | 0 | 9 | 0.0% | 0 | 30 | 0.0% |
| | Chaucer Melibee | 0 | 5 | 0.0% | 6 | 51 | 11.17% |
| | Chaucer Parson's Tale | 6 | 21 | 28.6% | 19 | 77 | 24.7% |
| | Edmund (Vernon ms.) | 0 | 9 | 0.0% | 2 | 15 | 13.3% |
| | Treatise on Horses | 0 | 2 | 0.0% | 1 | 3 | 33.3% |
| | Mandeville's Travels | 6 | 51 | 11.7% | 10 | 95 | 10.5% |
| | Wycliffe New Testament | 0 | 19 | 0.0% | 0 | 16 | 0.0% |
| | Wycliffe Old Testament | 0 | 2 | 0.0% | 0 | 0 | – |
| | Polchronicon | 0 | 10 | 0.0% | 0 | 37 | 0.0% |
| | Purvey Bible Prologue | 0 | 31 | 0.0% | 1 | 25 | 4.0% |
| | Wycliffe Sermons | 0 | 47 | 0.0% | 0 | 84 | 0.0% |
| | Total | 18 | 327 | 5.5% | 63 | 599 | 10.5% |

Brut (West Midlands). The Rule of St. Benet (c.1425, North) provides the most robust evidence of asymmetric negative concord in this period, as the examples in (154) illustrate.

(154) a. Na þing sal faile þaim þat wil luue god.
No thing shall fail them that will love God.
'Nothing shall fail those who love God'
(CMBENRUL,7.208)

[6] This group comprises: *Sawles Warde, Hali Meidhad, St. Katherine, St. Juliana, St. Margaret.*

b. he ne wyl noght lere na sience
he NEG will not learn no science
'he will learn no science'
(CMBENRUL,39.1262)

Numbers of examples from other texts are too small to discern a pattern either of strict or non-strict NC. Despite the scarcity of data from prose texts, what few prose data we have are broadly in accord with Ingham's (2006) verse data, showing a tripartite dialect split in Middle English negative doubling. Until the fourteenth century, Southern and South-Western texts exhibit strict negative doubling, some East Midlands texts exhibit non-strict negative doubling, and northern texts tend to lack negative doubling altogether.[7] The system of negation in these northern texts is similar to Present-day Standard English, in which each negative item has negative force and therefore there are no concordant negative items.

### *7.3.2 Variation in the Quantificational Semantics of Concordant Negatives*

Many accounts of negative concord, beginning with Ladusaw (1992), treat concordant negative words as indefinites. However, the behaviour of Old English negative objects indicates that negative words are quantificational. Typically, when a clause contains a non-finite verb within VP, negative objects, as in (155), and quantified objects, as in (156), precede the non-finite verb more frequently than other types of objects do.

(155) man ne dorste nan þing ogean his willan don
Men NEG dared no thing against his will do
'Men dared not do anything against his will'
(cochronE,ChronE_[Plummer]:1086.85.2904)

(156) he mæig ealle þing gefremman
he may all things accomplish
'he may accomplish all things'
(cocathom1,+ACHom_I,_33:462.102.6637)

(157) heo ne mage bearn habban
she NEG may-SUBJ children have
'she cannot have children'
(coaelhom,+AHom_20:106.2989)

While Pintzuk (2000, 2005), Pintzuk and Taylor (2003), Taylor and Pintzuk (2012) argue that the headedness of VP varies in Old English, and therefore that some of these instances of OV order involve head-final VPs, they argue that not all of the OV word orders we find in OE and ME can be derived using a head-final VP structure. The clearest evidence for this

[7] The use of negative concord in the northern dialect, is not unknown, however, and is subject to variation. The Rule of St. Benet is another northern text. While negative concord is not typical of this text, it does exhibit a small amount of asymmetric negative concord.

are examples which involve elements occurring to the right of the finite verb which cannot be postposed to that position, such as prepositions, particles as in (158), or pronouns as in (159). These provide unambiguous evidence for leftward movement of an object out of a head-initial VP – the pronouns and particles must be in their base-positions following the verb.

(158) þu ne mihtst nænne weg findan ofer
you NEG can no way find across
'you cannot find a way across'
(coboeth,Bo:34.85.22.1633)

(159) he ne mai nan þing don us buten godes leaue
he NEG may no thing do us but God's leave
'he may do nothing to us without God's leave'
(CMANCRIW,II.169.2346)

Pintzuk (2000, 17) claims that object movement to a position preceding a non-finite verb is almost entirely limited to quantified and negative objects in Old English. She estimates the frequency of non-negative, non-quantified object movement as at most 5%. Similarly, Kroch and Taylor (2000, 162) suggest that non-quantified, non-negative objects 'move leftwards at a very low rate, if at all' in Early Middle English. Thus the parallels between negated and quantified objects in terms of object movement are consistent with an analysis in which OE and ME concordant negatives are quantified NPs rather than indefinites.

Giannakidou (2000) argues that the difference between strict and non-strict negative concord can be reduced to a difference in the quantificational semantics of negative elements. In non-strict negative concord dialects, she hypothesises that concordant negative words are universal quantifiers ($\forall x$), whereas in strict negative concord dialects they are existentials ($\exists x$). This accounts for their different positions relative to negative markers. $\forall x \neg$ and $\neg\exists x$ are logically equivalent, but existential and universal quantifiers stand in a different relationship to negation. To achieve the appropriate reading relative to negation, existential quantifiers must stand within the scope of negation (there does not exist one x such that the proposition P is true of x), as in non-strict negative doubling where a negative marker C-commands the concordant negative word. On the other hand, universal quantifiers must scope over the negation (for every x, the proposition P is not true), either overtly or covertly by the time LF is reached. This option is overtly instantiated in strict negative doubling when the concordant negative word C-commands the negative marker. Thus the distinction between strict and non-strict negative doubling dialects reduces to a difference in the quantificational semantics of concordant negative items.

In strict negative doubling dialects of OE and ME, as in Greek (cf. Giannakidou (2006)), concordant negative objects can topicalise. When they do, they have a universal quantifier interpretation ($\forall x, \neg P(x)$), presupposing the existence of a discourse-given set of individuals to which the predicate does not apply. In examples like (160), the concordant negative introduces an existential presupposition. It is discourse-given that these sets are non-empty.

(160) a. ne for noon harm that men doon or seyn, he ne
nor for no harm that men do or say, he NEG
eschawfeth nat agayns his resoun.
become inflamed not against his reason
'Despite all the hurtful things men said or did, he did not become angry against his reason'
$(\forall(x), \text{harm }(x) \wedge \neg\text{inflame }(x, him))$
(CMCTPARS,310.C1.924)

b. This Gowanus broþer & his folk þat were Sarasynes
'This Gowanus' brother and his people that were Saracens
went þrouȝ-oute þe Lande and destroyede al þing þat þai
went throuhgout the land and destroyed everything that they
fonde and no þing þai ne sparede
found and nothing they NEG spared
'This Gowanus' brother and his people, the Saracens, went throughout the land destroying everything they found. They spared nothing.'
$(\forall(x), \text{thing they found}(x) \wedge \neg\text{spared}(x))$
(CMBRUT3,45.1352)

In strict negative doubling dialects, concordant negative subjects can also have this reading of existential presupposition when they C-command the negative marker. In (161), the subject *nan fur* 'no fire' is both discourse-old and existentially presupposed in the context – that is, a non-empty set.

(161) þe goldsmið fondeð þe gold i þe fur þe false gold for wurðeð
the goldsmith tests the gold in the fire and false gold destroys
þerin þe god brichtere kimeð out. sechenesse is anbrune hat to
therein the good brighter comes out. sickness is a flame hot to
þolien ac nan fur ne clenseð þe gold as hit deð þe sawle
suffer but no fire NEG cleanses the gold as it does the soul
'The goldmsith tests the gold in the fire and destroys the false gold in it so the good comes out brighter. Sickness is a hot flame to suffer but no fire cleanses gold so well as it cleanses the soul'
(CMANCRIW,II.138.1850)

By contrast, negative existentials $(\neg\exists x, P(x))$ do not introduce existential presuppositions – the set of individuals $x$ could be empty. That strict negative doubling involves universal quantifier concordant negative words will become crucial to the analysis presented in Chapter 8, and to our understanding of the relationship between the Jespersen Cycle and negative concord.

While this variability in the quantificational semantics of concordant negative words can explain almost all the variation we see in Old English negative doubling, it does not explain the loss of negative doubling in Middle English (both strict and non-strict negative doubling – see Table 7.3). Chapter 8 addresses this change, and its relationship to the Jespersen Cycle, arguing that the loss of negative doubling begins as a change in the way negative force is marked in negative doubling constructions concurrent with the same change in sentential negation contexts.

## 7.4 Negative Spread in Early English

In addition to negative doubling, we find instances of negative spread in early English like those in (162). These instances of negative concord lack a negative marker.

(162) no man seyd no-thyng a-gens hem
no man said nothing against them
'no man said anything against them' (CMKEMPE,33.730)

Examples like (162) appear sporadically in corpus data from the ninth century until the fourteenth century, but are most frequent from the fourteenth century until the late sixteenth century. There are 111 instances in the PPCME2, mostly dating from the late fourteenth and fifteenth centuries after the loss of negative doubling with *ne*. There are also twelve examples in the PPCEME, dating from the sixteenth century. By contrast, there are only four instances in the YCOE. Negative spread increases in frequency during the fourteenth century due to the loss of *ne*.

Chapters 8 and 9 argue that the loss of *ne* proceeds in two stages in negative concord environments just as it does in the Jespersen Cycle – loss of $ne_1$ and loss of $ne_2$. From the thirteenth century onwards, after the loss of $ne_1$ I will argue that most Middle English negative concord is in fact negative spread, rather than negative doubling. Negative doubling requires the negative marker to contribute a negative operator $\neg$ at LF. This negative operator ($ne_1$ or *not*) contributes negative force to the clause and licenses concordant negative words. It follows that when ME negative concord involves the negative marker $ne_2$, it is not negative doubling, because $ne_2$ lacks negative force, and is not a negative operator. In this respect ME negative concord is like Present-day French negative concord.

Giannakidou (2006, 24) argues that in Present-day French negative concord, the negative marker *ne* is not a negative operator but an optional concordant negative item licensed by negative spread. In negative spread, the structurally highest negative word is the negative operator, contributing negative force to the clause and licensing concordant negative items in its scope, including *ne*. Chapter 8 identifies environments in which negative doubling with $ne_1$ [iNeg] and negative spread with $ne_2$ [uNeg] are syntactically distinct, and in which we can therefore observe competition between $ne_1$ and $ne_2$ in Late Old English and Early Middle English. In Chapter 9, I argue that this distinction between negative doubling and negative spread is crucial to understanding the loss of negative concord.

# 8 Negative Inversion

## *Evidence for a Quantifier Cycle in Early English*

### 8.1 Negative Inversion and the Quantifier Cycle

In this chapter, I turn to the question of which negative words in negative concord have negative force. Feature-based accounts of negative concord, for example Zeijlstra (2004), argue that only one negative word in concord bears negative force and is interpretable as a negative operator at LF. This chapter considers evidence for variation and change in the negative force associated with negative arguments and negative adverbials in early English, then examines how this variation and change interacts with the Jespersen Cycle in negative concord environments.

In negative concord of the type in (163), it is not clear which negative item contributes negation at LF, and which is simply negative in form.

(163) Ic ne funde nanne gylt on him
I NEG found no sin in him
'I found no sin in him'
(cowsgosp,Jn_[WSCp]:19.6.7261)

Is it *ne* or the negative quantifier *nanne* 'no' that marks negative force in (163), or is negative force marked by a non-overt operator (as Zeijlstra (2004) argues)? If the negative marker *ne* marks negative force and *nanne* 'no' is a concordant negative word, then (163) is an instance of negative doubling. Similarly, if a null negative operator marks negative force in (163) then it involves negative doubling between the null operator and the concordant negative words *ne* and *nanne* 'no'. However, if the negative quantifier *nanne* 'no' marks negative force and *ne* is in concord with it, then (163) exemplifies negative spread, of the type observed in Present-day French by Giannakidou (2006) and Zeijlstra (2010).

Patterns of negative inversion are not widely discussed in the literature (but see Wallage (2012a), Ingham (2007), Nevalainen (1997)). Here I argue they are a diagnostic that allows us to distinguish words that mark negative force at LF from those that are negative in form only. Changing patterns of negative inversion in OE and ME demonstrate that there is a quantifier cycle of the type proposed by Ladusaw (1993) – a change in the semantic representation of negative words in negative concord – that is bound up with the morphosyntactic weakening of *ne*. There is competition between two types of negative concord at successive stages of the quantifier cycle – negative doubling, in which $ne_1$ marks negative force at LF; and negative spread, in which $ne_2$ does not.

Recall from Chapter 4 that negative inversion functions as polarity focus. Therefore the initial negative must be a negative operator at LF. Haegeman (2001) observes for Present-day English that only initial negative elements with sentential scope are triggers for subject-operator inversion. In (164a) the negative in spec,CP has sentential scope, and there is subject-operator inversion (T to C movement). By contrast, in (164b), the negation has constituent rather than sentential scope, so there is no inversion.

(164) a. With no job would Mary be happy
(=Mary wouldn't be happy with any job)
(Haegeman 2001: 21, ex.1a)
b. With no job, Mary would be happy
(=Mary would be happy without a job)
(Haegeman 2001: 21, ex.1b)

Extending this idea to earlier periods of English, the emergence of new contexts for negative inversion provides evidence that the semantic properties of clause-initial negative items change in negative concord environments. Ladusaw (1993) argues that the development of negative arguments and adverbials follows the stages of what he terms an argument cycle or quantifier cycle (Ladusaw, 1993, 437) outlined below:

> The development of negation-expressing argument phrases from regular indefinite arguments has the following stages: first the argument is a regular indefinite argument, then it becomes a co-occurring supporter of the clausal negation, and finally it becomes an independent expressor of negation. (Ladusaw, 1993, 437–438)

Ladusaw (1993, 438) proposes a four stage model of the quantifier cycle, with the stages as set out in (165):

(165) a. **Stage One:** Argument (not polarity sensitive): *She didn't say one thing* (cf. its affirmative equivalent *She said one thing*)
b. **Stage Two:** Non-veridical polarity item: *She didn't say anything*
c. **Stage Three:** Concordant negative item or anti-veridical polarity item: *She didn't say nothing*[1]
d. **Stage Four:** Negative (syntactically autonomous and expresses negative force at LF): *She said nothing*

Each stage of the quantifier cycle has been attested in various languages, for example German (Jaeger, 2008) and Welsh (Willis, 2011b). Willis (2011b)

[1] Willis (2011b) suggests that some arguments in some languages are sensitive to a distinction between anti-morphic (occurring only with a sentential negative marker) and anti-veridical (occurring with any element that bears negative force).

and van der Auwera and Alsenoy (2011) provide good overviews of the cross-linguistic data. However, only the final two stages are relevant for our purposes here.

Once we allow variation between negative items that are semantically interpretable (negative operators at LF) and those that are semantically uninterpretable (negative in form only), new patterns of negative inversion result as concordant negative words are reanalysed as negative operators. Hence negative inversion provides evidence for a transition between stage three (165c) and stage four (165d) of the quantifier cycle. Only negative quantifiers with negative force at LF trigger subject–verb inversion when they appear in clause-initial position. Concordant negative indefinites cannot.

## 8.2 Subject–verb Inversion in Old and Middle English

Section 4.3 described the syntax of the various types of subject–verb inversion available in Old and Middle English. To recap, inversion of nominal subjects in spec,TP and finite verbs is found in a wide range of clauses, including affirmative declaratives. Haeberli (2002a) analyses this as V to Agr movement in main clauses.[2] Subject pronouns appear in spec,AgrP. Hence, inversion of a finite verb and subject pronoun indicates verb movement to $C^0$ in the clause structure (166). V to C movement is typical of certain clause types in Old English and Middle English, including interrogatives, imperatives, clauses with initial discourse adverbs, subjunctives, conditionals and negatives.

(166)

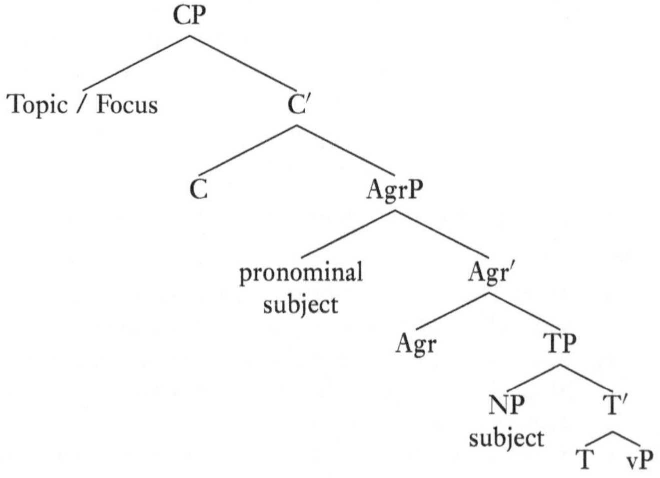


2 Though variation should be noted here. Kroch and Taylor (1997) argue that there is a generalised pattern of V to C movement in northern Middle English (texts such as the *Rule of St Benet*) that occurs even in affirmative declaratives. This is similar to verb-second in languages such as German. Haeberli (2002b) argues that there is much more variation between V to C movement and V to Agr movement in affirmative declaratives than hitherto assumed, even in East Midlands texts.

Given that inversion with nominal and pronominal subjects may represent verb-movement to different positions in Old and Middle English, we should restrict the analysis to clauses in which V to C movement can be identified unambiguously – that is clauses with demonstrative or personal pronoun subjects. Furthermore, as imperatives, subjunctives and conditionals trigger V to C movement irrespective of the polarity of the clause, the analysis of negative inversion must be restricted to clauses with unambiguously morphologically marked indicative verbs. Examples like (167) which involve a clause-initial negative but are also subjunctive or imperative in mood are therefore excluded from the analysis.

(167) Ne þenc þu be mergene
NEG think you about morrow
'Don't think about tomorrow.'
(ÆLS 31.57, Pintzuk (1999, 91, ex.115)

## 8.3 The Diachrony of Negative Inversion

Throughout the history of English, declarative main clauses with clause-initial negatives are a context for subject–verb inversion. However, the clause-initial negative elements that trigger inversion change over time. In Old English and Early Middle English, subject–verb inversion follows clause-initial *ne* as illustrated in (168), whereas in later periods, subject–verb inversion follows clause initial negative quantifiers such as *no* (169), adverbials such as *never* (170) and negative conjunctions such as *nor* (171).

(168) Ne hafst tu næure soðe eadmodnesse on þe...
NEG have you never true humility in you...
'You never have true humility in you...'
(13th century; VICES1,33.398)

(169) and no wepyn coude he fynde
and no weapon could he find
'and no weapon could he find'
(15th century; MALORY,64.2150)

(170) never took he giftes of man
never took he gifts of man
'Never did he take gifts of man'
(15th century; CAPHR,54.666)

(171) Nor did he use those obligations ill, that love had done him
(17th century; BEHN-E3-H,156.13)

Table 8.1 shows the frequency of three types of negative inversion in early English. These are clauses with initial *ne*, as in (168); clauses with initial negative arguments or adverbials, as in (169) and (170); and clauses with initial negative conjunctions, as in (171). The Old English data (850-1150) come from the YCOE (Taylor et al., 2002), the Middle English data (1150–1500) from the PPCME2 (Kroch and Taylor, 2000) and the Early

Table 8.1. *The frequency of inversion following different clause-initial negative items (excluding clear cases of constituent negation and verbs that are morphologically marked subjunctive or imperative).*

| | Ne | | | Neg argument & adverbial | | | Neg conjunction | | |
|---|---|---|---|---|---|---|---|---|---|
| Period | V su | Total | % V su | V su | Total | % V su | V su | Total | % V su |
| 850–950 | 148 | 255 | 58.0% | 2 | 6 | 33.3% | 0 | 35 | 0.0% |
| 950–1050 | 351 | 651 | 53.9% | 5 | 22 | 22.7% | 2 | 203 | 1.0% |
| 1050–1150 | 67 | 131 | 51.1% | 3 | 5 | 60.0% | 2 | 48 | 4.0% |
| 1150–1250 | 36 | 77 | 46.8% | 8 | 9 | 88.9% | 9 | 27 | 33.3% |
| 1250–1350 | 0 | 18 | – | 0 | 0 | – | 1 | 18 | 5.6% |
| 1350–1420 | 0 | 6 | – | 7 | 14 | 50.0% | 4 | 50 | 8.0% |
| 1420–1500 | 0 | 0 | – | 9 | 11 | 81.8% | 7 | 34 | 20.5% |
| 1500–1570 | – | – | – | 14 | 23 | 60.9% | 36 | 81 | 44.4% |
| 1570–1640 | – | – | – | 24 | 31 | 77.4% | 223 | 235 | 94.9% |
| 1640–1710 | – | – | – | 2 | 3 | 66.7% | 159 | 169 | 94.1% |

Modern English data (1500–1710) from the PPCEME (Kroch et al., 2004). The Late Middle English (fourteenth and fifteenth century) and Early Modern English data are supplemented by data from the PCEEC (Taylor et al., 2006). Where a text appears in both the PCEEC and the PPCEME, each example from that text is only counted once.

Earlier accounts of negative inversion such as Nevalainen (1997) and van Kemenade (2001) describe a gap in the productivity of negative inversion during the Middle English period, between the loss of the Old English pattern with *ne* in the thirteenth century and the development of the Modern English pattern with negative arguments and adverbials, which both Ingham (2007) and Nevalainen (1997) claim emerges in the fifteenth and sixteenth centuries. However, data from the YCOE and PPCME2 provide evidence of inversion following negative arguments or adverbials in both Old English and Middle English. Table 8.1 demonstrates continuity in the availability of inversion: as inversion following *ne* declines in frequency in Early Middle English, inversion following negative arguments and adverbials increases in frequency. At no point is negative inversion unproductive.[3]

The negative conjunctions *nor, neither* do not typically become triggers for inversion until the seventeenth century, much later than negative arguments or adverbials do. Different patterns of inversion following negative arguments and adverbials on the one hand, and negative conjunctions on the other support Nevalainen's (1997) hypothesis that the property of triggering negative inversion is encoded on individual lexical items, such that negative conjunctions acquire the property later than negative arguments or negative adverbials. An account in which negative force is specified on individual

[3] It follows that the mechanism for movement of head *ne*, or phrasal negative arguments and adverbs to clause-initial position within CP is the same. It is motivated by the presence of an LF-interpretable [iNeg] feature that can be focused.

lexical items via their morphosyntactic feature specification easily accommodates such lexical variability.

## 8.4 The Relationship between Negative Inversion and Negative Concord

If the rise of inversion following clause initial negative arguments and adjuncts results from them becoming negative operators at stage four of the quantifier cycle, these data raise questions concerning the relationship between negative inversion, negative concord and the Jespersen Cycle.

Both Ingham (2007) and Nevalainen (1997) argue that negative inversion following a negative argument or adverbial is incompatible with negative concord, and emerges as negative concord is lost in the sixteenth and seventeenth centuries. Ingham (2007) proposes to derive the incompatibility of negative inversion and negative concord by hypothesising that negative concord between a negative operator (such as *not*) in spec,NegP and a negative item merged within VP blocks movement of the negative item to spec,CP across the operator in spec,NegP. In (172), there is an operator–variable dependency between the negative operator and the concordant negative indefinite object. Therefore, the negative operator creates an island out of which the negative indefinite cannot move.

(172)

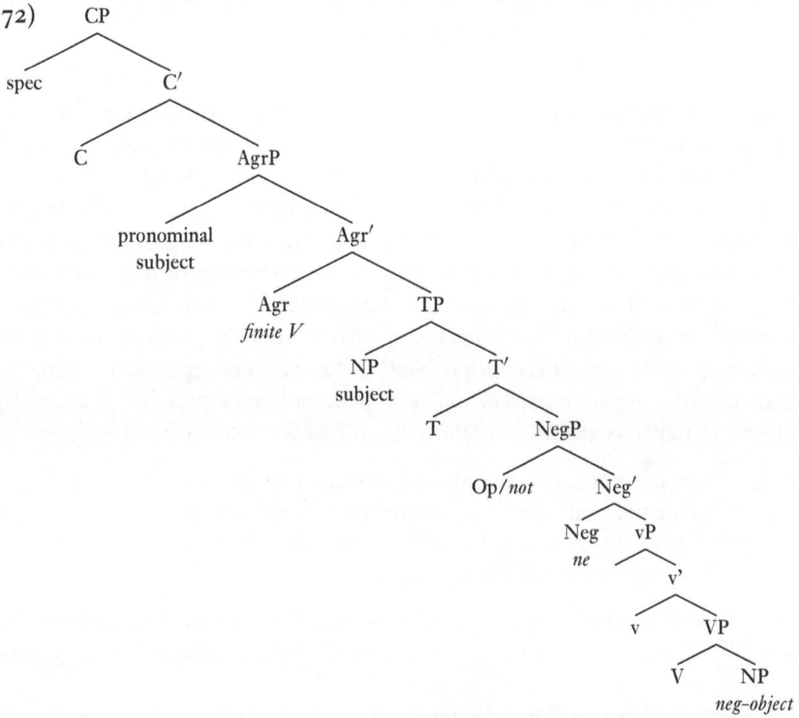


In earlier periods, where we find negative concord between *ne* and a negative item merged within VP, Ingham claims that this operator remains present in spec,NegP but that it is phonologically null. This account predicts not only that negative inversion and negative concord are incompatible, but that negative arguments and adverbials cannot escape NegP, so in clauses with negative concord, negative arguments or adverbials cannot appear in spec,CP or spec, TP. Hence Ingham's (2007) account predicts that early English negative concord is non-strict negative concord. This is not the case for all dialects of early English (Section 7.3).

The rise of negative inversion following negative arguments and adverbials occurs before the loss of negative concord itself. Clause-initial negative arguments and negative adverbials begin to trigger inversion during the thirteenth century while negative concord is still highly productive. In the Middle English period as a whole (1150–1500), 40.0% (n=12/30) of clauses in which inversion follows a clause-initial negative argument or adverbial also have negative doubling. (173) illustrates that negative inversion can follow a clause-initial negative argument or adverbial even when there is concord between a clause-initial negative argument or adverbial and *ne*. (173) involves negative concord between the clause-initial constituent *Nouther hwit ne blac* 'neither white nor black' and *ne*.

(173) Nouther hwit ne blac ne nemmet he in his ordre
Neither white nor black NEG names he in his order
'He does not name in his order either white or black'
(13th century; ANCRIW,I.48.84)

Inversion only follows negatives which contribute negative force at LF.[4] Therefore the derivation of a clause like (173) with both inversion following an initial negative argument and negative concord with *ne* requires that the clause-initial negative word bears negative force, and *ne* lacks it.

In (173) and (174), both from the thirteenth century, the different triggers for inversion in these negative concord constructions suggest different negative elements are semantically interpretable as markers of negative force at LF, and provide another argument for distinguishing two forms of negative concord: negative doubling with the negative marker $ne_1$ [iNeg] and negative spread with the concordant $ne_2$ [uNeg]. (174) exhibits negative doubling: the clause-initial negative marker $ne_1$ is semantically interpretable as a negative operator, negative inversion follows it, and *næure* is concordant with it.

(174) Ne hafst tu næure soðe eadmodnesse on þe...
NEG have you never true humility in you...
'You never have true humility in you...'
(13th century; VICES1,33.398)

On the other hand, (173) exemplifies negative spread: the clause-initial negative quantifier, rather than the negative marker, is the negative operator,

4 Although the converse, that initial negative items which negate a clause on their own (do not appear in negative doubling) always trigger negative inversion, is not true. It depends on their scope.

and negative inversion follows it. It licenses the concordant negative $ne_2$ in its scope.

The examples in (175) without negative inversion exhibit negative doubling: $ne_1$ is the marker of negative force. The clause-initial negative quantifiers in (175) are concordant items, negative in form only, hence they do not trigger inversion. They are universal quantifiers of the type discussed in Section 7.3.2.

(175) a. and in no maner a wyse he ne myght se that blessyd
and in no manner of ways he NEG might see that blessed
sacrament;
sacrament;
'and in no way might he see that blessed sacrament'
(CMGREGOR,234.2509)

b. nænne he ne fordemde
none he NEG judged
'he judged none'
(10th/11th century; aelive,+ALS_[Martin]:302.6153)

Negative doubling and negative spread are distinct in OE and ME: negative doubling and negative inversion are mutually exclusive, except when the negative marker is itself in clause-initial position as in (174); however, negative spread and negative inversion are compatible. Changes to negative inversion show how competition between these two types of negative concord progresses over time and interacts with the Jespersen Cycle. Table 8.2 reports the frequency of negative inversion in negative concord with *ne* and *not*.

In Table 8.2, negative inversion is much more frequent with *ne* than with *not*. The frequency of negative inversion in negative concord with *ne* increases during the transition from Old English to Middle English. Indeed, the frequency of negative inversion in ME is similar irrespective of whether the clause exhibits negative concord with *ne*, as in (176); or whether the clause-initial negative is the only negative item in the clause, as in (177).[5]

[5] However, irrespective of the presence or absence of negative doubling, there are some clauses with clause-initial negatives which lack inversion, such as those in i.-iii.

i. and neuer þai rest til þat þai hade here taken, and put her vnto deth.
and never they rest until that they had her taken, and put her to death.
'and they never rested until they had taken her and put her to death.'
(CMBRUT3,21.600)

ii. but neuer þai haden childe to-gedres,
but never they had a child together
'but they never had a child together'
(CMBRUT3,77.2355)

iii. and ȝitt neuer He sayde till thaym anes why þay swa dyde.
and yet never He said to them once why they so did.
'and yet he never once told them why they did so.'
(CMEDTHOR,43.622)

Table 8.2. *The interaction of negative inversion and negative doubling according to negative marker. Clauses with initial negative arguments, adverbials or conjunctions only.*

| | with *ne* | | | with *not* | | | w/o neg concord | | |
|---|---|---|---|---|---|---|---|---|---|
| Period | inv | Total | % inv | inv | Total | % inv | inv | Total | % inv |
| Old English[6] (850–1150) | 6 | 22 | 27.3% | 0 | 0 | – | 0 | 3[7] | 0% |
| Middle English (1150–1500) | 11 | 13 | 84.6% | 1 | 3 | 33.3% | 18 | 21 | 85.7% |
| Early Modern (1500–1710) | 0 | 0 | – | 3 | 39 | 7.7% | 455 | 484 | 94.0% |

(176) and nothing ne shal they fynden in hir handes of all hir
and nothing NEG shall they find in their hands of all their
tresor
treasure
'and they shall find none of their treasure in their hands'
(14th century; CTPARS,292.C1.156)

(177) and no defaute fond þei in hir feith...
and no fault found they in her faith...
'and they found no fault in her faith'
(15th century; CAPCHR,159.3739)

These are all second conjuncts, in which it seems negative inversion remains variable. If we exclude second conjuncts from the figures for the Middle English period, negative inversion becomes categorical (n=5/5 in negative concord with *ne*, n=5/5 in clauses without negative concord).

6 Excluding all instances of morphological subjunctives, or morphological imperatives, which typically have V to C movement irrespective of their polarity.

7 These occur only in the Old English *Bede*, and are given as i.-iii.

i. & cwæð: Næfre ofer þis ic owiht ma spreco
& said: Never of this I anything more say
' (cobede,Bede_3:12.196.26.1994)

ii. næfre he ða his wætan hræl & þa cealdan forlæatan wolde
never he then his wet garments and the cold lay aside would
'he never would lay aside his wet and cold garments...'
(cobede,Bede_5:13.436.3.4371)

iii. æfter þon heo cwæð: Nænge þinga ic þas bliðe aberan mæg.
after then she said: No thing I this gladly suffer may
'she said afterwards: I am not at all willing to suffer that'
(cobede,Bede_4:12.290.18.2930)

It is unclear how to interpret these examples, or evaluate their significance as counter-examples. This particular text is a close translation of Bede's Latin original, and is often remarked upon for its unnaturalness (Rowley, 2011). Waite (2013) also remarks on the translation's unidiomatic syntax, and Whitelock (1962) comments that the translator uses Latinate constructions and an unidiomatic word order. It is clear that they do not represent a pattern productive in any other Old English text.

In both (176) and (177) the clause-initial negative argument is the negative operator. The only difference between them is that (176) exhibits negative spread – the clause-initial negative operator licenses concordant *ne*2 in its scope – whereas in (177) there is no *ne*2.

Conversely, Middle English *not* behaves like Old English *ne* with respect to negative inversion. This is expected if both are negative operators. Negative inversion and *not* are different ways to mark sentential scope negation, and as such should be mutually exclusive. In fact, they co-occur very infrequently during Middle English and Early Modern English, occurring in only 9.5 per cent (n=4/42) of clauses with clause-initial negative elements like (178). An analysis of these four exceptional cases has to motivate the inversion in a way that is independent of negation.[8]

(178) a. and for non oþer þing is it not good
and for no other thing is it not good
'and it would not be good for anything else' (15th century; CMMANDEV,83.2101)
b. no more would not I if I was your wife 'No more would I if I was your wife'
(17th century; PENNY-E3-H,267.501)

## 8.5 The Quantifier Cycle and the Jespersen Cycle

By distinguishing two forms of *ne* at successive stages of the quantifier cycle, just as we did for the Jespersen Cycle in Chapter 5, changes in the availability of negative inversion and the form of negative concord follow from the interaction of the quantifier and Jespersen Cycles in ME. The incompatibility between negative inversion and negative doubling noted by Ingham (2007) and Nevalainen (1997) for Early Modern English is really an incompatibility between negative inversion and negative doubling – negative concord involving the negative operators *ne*1 or *not*, as illustrated in (179) and (180).

(179) nan þing he ne answarode
no thing he NEG answered
'he did not give any answer'
(11th century; wsgosp,Mt_[WSCp]:27.12.2014)

(180) a. for in none oþere maner þai mygh nouȝt hym avenge.
for in no other manner they might not him avenge
'for they could not avenge him in any other way'
(14th century, CMBRUT3,125.3794)

8 (178b) typifies the 3 Early Modern English examples. All are clearly conditional, which might account for inversion in these examples. (178a) is also an irrealis context (although not marked by the morphological subjunctive) expressing a judgement. This might motivate V to C movement irrespective of polarity.

b. ne for noon harm that men doon or seyn, he ne
nor for no harm that men do or say, he NEG
eschawfeth nat agayns his resoun.
become inflamed not against his reason
'he did not become angry against his judgement, no matter what hurtful things men said or did'
(14th century, CMCTPARS,310.C1.924)

c. but by no meanes she would not confesse the same
but by no means she would not confess the same
'but by no means would she confess the same'
(16th century; ORIGIN2,287.030.461, PCEEC)

d. nor I do not care
'nor do I care'
(17th century; LISLE-E3-H,IV,123C1.502)

Patterns of negative inversion distinguish two types of negative concord: negative doubling in which the negative marker (*ne*$_1$ or *not*) has negative force and licenses its clausemate negative arguments and adverbials, as in (181a); and negative spread in which *ne*$_2$ lacks negative force and is itself licensed by a negative argument or negative adverbial that marks negative force, as in (181b).[9]

(181) a. nænne he ne fordemde
none he NEG judged
'he judged none'
(10th/11th century; aelive,+ALS_[Martin]:302.6153)

b. Nouther hwit ne blac ne nemmet he in his ordre
Neither white nor black NEG names he in his order
'He does not name in his order either white or black'
(13th century; ANCRIW,I.48.84)

The inversion data I present here are further evidence that semantic weakening of *ne* precedes loss of the morpheme itself, not only in sentential negation contexts, but in negative concord too. Ingham (2011) concludes on the basis of Anglo-Norman data that Middle English concordant negatives have become – or are becoming – negative quantifiers considerably before the fourteenth-century loss of *ne*. He claims that there is no reliable Middle English prose evidence for this development. However, the negative inversion data presented here date this development to the twelfth and thirteenth centuries, in prose texts.

9 A welcome consequence is loss of the distinction between strict and non-strict negative doubling in early ME as negative doubling is replaced by negative spread. The data in Table 7.3 (Chapter 7) provide much less evidence of a strict–non-strict distinction in ME than in OE. This follows if much ME negative concord is not negative doubling but negative spread.

While inversion shows that some negative arguments and adverbials become negative quantifiers at this early date, not all do. When negative arguments co-occur in negative spread, as in (182), only the structurally highest negative argument is an operator.

(182) no man seyd no-thyng a-geyns hem
no man said nothing against him
'no man said anything against him' (CMKEMPE,33.730)

The lower one is a concordant negative item (representing stage three of the quantifier cycle), licensed in the scope of the higher one. While negative spread remains productive, two homophonous sets of negatives co-exist, negative operators and concordant negative items.

## 8.6 The Syntactic Representations of Early English Negative Concord

The negative inversion data in Section 8.4 have several implications for an analysis of negative concord in early English, and suggest certain principles that should shape the analysis. The licensing relation in negative concord should be same as in Jespersen Cycle – involving the same elements and relationships between them. This allows us to formalise the interaction between negative concord and the Jespersen Cycle revealed by the negative inversion data. In negative doubling, as in sentential negation, we need to distinguish two forms of *ne*: a negative operator $ne_1$ [iNeg] and a concordant negative item $ne_2$ [uNeg]. When it occurs in negative doubling, from the fourteenth century onwards, *not* behaves as a negative operator with an LF-interpretable [iNeg] feature. Turning to negative arguments and adverbials, I hypothesise that their feature-specifications change over time. I hypothesise that they have a [uNeg] feature at stage three of the quantifier cycle, and become negative operators at stage four when they gain an [iNeg] feature. Given an appropriate mechanism of syntactic agreement, the changes in negative concord and sentential negation contexts – and the relationship between them – emerge out of the interaction between morphosyntactic feature-based analyses of both the Jespersen cycle and the quantifier cycle.

All uninterpretable [*u*Neg] features are formal only, and need to be checked and deleted prior to interpretation at LF. Their function is to create dependencies between items that are negative in form and negative in semantic interpretation. The relationships between features should be those which hold between morphosyntactic features more generally – agreement relations between a probe and a goal (Chomsky, 1999, 2000). In Section 5.4, I argued that the syntactic analysis of Jespersen Cycle required the mechanism of Reverse Agree proposed by Wurmbrand (2012), Merchant (2013) and Zeijlstra (2012), defined by Zeijlstra (2012, 17) as in (183).

(183) Reverse Agree: $\alpha$ can Agree with $\beta$ iff:

a. $\alpha$ carries at least one uninterpretable feature and $\beta$ carries a matching interpretable feature.

b. $\beta$ c-commands $\alpha$.

c. $\beta$ is the closest goal to $\alpha$.

(Zeijlstra, 2012, 17)

Under (183), the uninterpretable formal features on negative items [*u*Neg] are checked by multiple agreement during the syntactic derivation, by a negative which has a semantically interpretable [*i*Neg] negative feature. This deletes all the uninterpretable [*u*Neg] features, leaving only the interpretable ones at LF. Multiple agreement falls out of reverse or upwards agree provided that the [iNeg] feature C-commands all the [uNeg] features. Each [uNeg] feature will probe upwards until it finds a matching [iNeg] feature, and in principle each [uNeg] feature (probe) could have the same [iNeg] feature goal. Reverse agree derives negative doubling at each stage of the Jespersen Cycle, and also negative spread.

### *8.6.1 Non-strict Negative Doubling*

Turning first to non-strict negative concord, the data presented by Ingham (2006) and discussed in Chapter 7 suggest the availability of strict and non-strict negative concord is a matter of dialect variation in Old English for which the analysis must account. Non-strict negative doubling takes the forms in (184), at successive stages of the Jespersen Cycle.

(184) **Stage 1** Subject – ne+V [iNeg] … negative item [uNeg]
**Stage 2** Subject – ne+V [iNeg] – not [uNeg] … negative item [uNeg]
**Stage 3** Subject – ne+V [uNeg] – not [iNeg] … negative item [uNeg]
**Stage 4** Subject – V – not [iNeg] … negative item [uNeg]

In non-strict negative doubling, Giannakidou (2000) argues negative items are existential quantifiers within the scope of the LF-interpretable negative marker (negative operator). The negative marker must C-command the negative indefinites for the appropriate negative concord reading ($\neg\exists x$) to hold.

A sample derivation for negative doubling at stage four of the Jespersen Cycle (185) is shown in (186).[10] [*u*Neg] on the negative head $Neg^0$ and on the DP *no sale* are the probes and [iNeg] on *not* the goal.[11]

[10] In (185) and all the derivations that follow, deleted features are distinguished by the use of strike-through.

[11] For ease of exposition, I assume that the modal *might* is merged in $T^0$. Nothing within the analysis follows from this.

(185) we myght nott make no sayle in Cristmasse wek
we might not make no sale in Christmas week
'We might not make any sale in Christmas week'
(16th century; TORKINGT-E1-H,58.328)

(186)

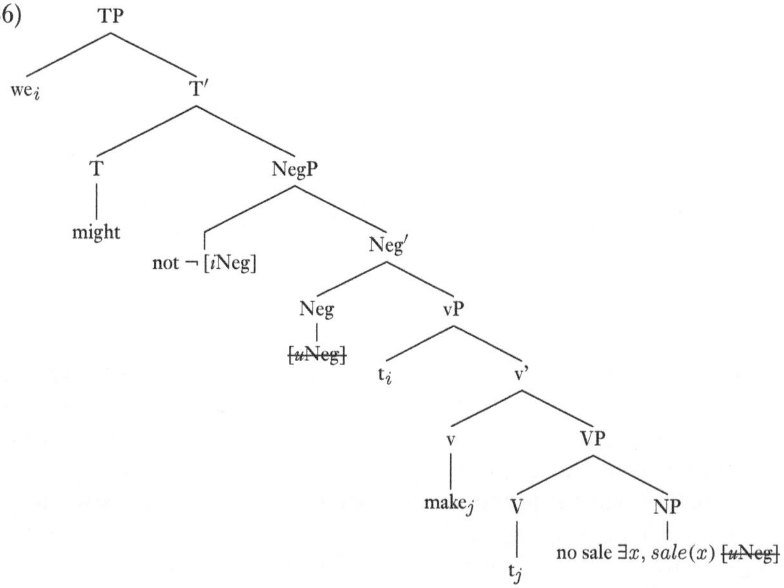


An example like (187) at stage three of the Jespersen Cycle follows a similar derivation, except that $Neg^0$ is overt *ne*.

(187) Thou ne shalt nat eek make no lesynges in thy confessioun...
You NEG shall not also make no falsehoods in your confession...
'You shall also not tell any lies in your confession. . . ' (14th century; CMCTPARS,325.C1a.1581)

At stages one and two of the Jespersen Cycle, the derivation is similar, except that *ne* is specified [iNeg] and *not* is therefore absent.

Negative concord in English is generally clause-bound, as in other languages (cf. Zeijlstra (2004) for crosslinguistic data). However, there is one exception. I showed in Chapter 3 that negative concord can also hold across clause boundaries in particular clause types – non-assertive clauses with irrealis interpretations like (188).

(188) I wuld not in no wyse ye shuld put your-self in no daungere
I intended not in no way you should put yourself in no danger
to hym
to him
'I did not intend that you should put yourself in any danger from him'
(PASTON,I,356.115.3554)

These non-assertive contexts can be marked by subjunctive verbs, although the morphological subjunctive is not always present. Syntactic agreement derives locality constraints on negative concord.[12] Giorgi (2004) argues that non-assertive subjunctive clauses in Italian are not syntactic islands, and suggests that they lack a force projection or feature in CP. Since they do not have force features of their own in CP, the force features of the subordinate clause do not intervene to block Reverse Agree between negative words in the superordinate and subordinate clause. Giorgi (2004) argues that these non-assertive CPs do not constitute phases in phase-based models of locality such as Chomsky (1999). So (188) simply has the agree relation in (189)

(189) [$_{CP}$ I wuld$_i$ [$_{NegP}$ not [iNeg] [$_{vP}$ in no ~~[uNeg]~~ wyse [$_{vP}$ t$_i$ [$_{CP}$ ye should [$_{vP}$ put yourself in no ~~[uNeg]~~ daungere to hym]]]]]]

### *8.6.2 Strict Negative Doubling*

In non-strict NC languages, concordant negative items [uNeg] must remain C-commanded by the negative marker [iNeg] at spellout. However, in strict NC languages, they can move to positions structurally higher than the negative marker. Examples of strict negative doubling where a negative word is not C-commanded by the negative marker such as those in (190) are problematic for a variable binding approach to negative concord, and demonstrate that negative words in strict negative concord are not indefinites.

(190) a. and no man myght not cope whythe hym, tylle the kynge prayd hym to be mery and sende hym a tokyn,
'and no man might cope with him, until the king prayed him to be merry and send him a token'
(15th century; CMGREGOR,219.2123)
b. but by no meanes she would not confesse the same
'but by no means would she confess the same'
(16th century; ORIGIN2,287.030.461, PCEEC[13])

Zeijlstra (2004) argues that concordant negative words are indefinites. He (2004, 245) links the availability of strict and non-strict negative concord to the syntactic features of the negative marker. He hypotheses that the negative markers in strict negative concord languages have [uNeg] features and that the negative operator is null in these languages, whereas the negative markers in non-strict negative concord languages have [iNeg] features and are morphologically overt. However, the availability of strict negative concord

[12] By adopting Reverse Agree rather than Agree, we can also explain why the negative item in the subordinate clause of examples like (188) is always concordant – it follows from the feature configuration required by Reverse Agree.

[13] This example comes from the Parsed Corpus of Early English Correspondence (Taylor et al., 2006).

does not vary across successive stages of the English Jespersen Cycle as his hypothesis predicts. Examples like (190) show that strict negative concord is possible in fifteenth- and sixteenth-century English with the morphologically overt negative operator *not* [iNeg]. Under Zeijlstra's analysis we are forced to analyse *not* as a concordant negative item (an anti-veridical NPI) when it appears in negative concord, despite the fact that elsewhere *not* is grammaticalised as the default sentential negative marker by the fifteenth century. Zeijlstra's account of strict negative doubling is incompatible with the analysis of the Jespersen Cycle I proposed in Chapter 5.

Zeijlstra's analysis does not predict the interaction between the Jespersen Cycle, strict negative doubling and negative inversion we find in early English either. Zeijlstra's (2004, 254) analysis of pre-verbal French n-words compounds a null negative operator onto the n-word as the n-word moves into subject position. Strict negative concord is negative spread, as he treats the negative marker in strict negative concord as a negative polarity item with a [uNeg] feature licensed by the pre-verbal argument+negative-operator compound [iNeg]. This analysis predicts that all pre-verbal n-words are in fact LF-interpretable negative operators. It follows that all clause-initial negative words in negative concord should trigger negative inversion. They are negative operators in a position of sentential scope. However, clause-initial negative words do not trigger negative inversion either with Late Middle English *not*, in (190b), or with Old English *ne*, in (191), precisely because $ne_1$ and *not* are themselves negative operators.

(191) nan þing he ne answarode
no thing he NEG answered
'he did not give any answer'
(11th century; wsgosp,Mt_[WSCp]:27.12.2014)

We have the four strict negative doubling patterns in (192) to account for at successive stages of the Jespersen Cycle.

(192) a. **Stage 1** Negative item [uNeg] – subject pronoun – ne+V [iNeg]
b. **Stage 2** Negative item [uNeg] – subject pronoun – ne+V [iNeg] – not [uNeg]
c. **Stage 3** Negative item [uNeg] – subject pronoun – ne+V [uNeg] – not [iNeg] (Jespersen Cycle)
d. **Stage 4** Negative item [uNeg] – subject pronoun – V – not [iNeg]

The link between the availability of negative inversion and the Jespersen Cycle is impossible to formulate under Zeijlstra's analysis, because his account rules out all the patterns in (192). To derive them, a concordant negative item must move across a morphologically overt LF-interpretable negative operator to a structurally higher position (for example to spec,TP, to clause-initial spec,CP or to a CP-adjoined position), having already had its [uNeg] feature checked under Reverse Agree by a C-commanding negative operator (with the feature [iNeg]) at an earlier stage of the derivation. Therefore concordant negatives in strict negative concord cannot be indefinites bound by an operator at S-structure.

This ability to move follows if concordant negatives that C-command negative operators are quantificational, as Giannakidou (2000, 2006) argues, rather than indefinites which must be bound under existential closure by a negative operator at spellout (S-structure). If concordant negatives in strict NC languages can be universal quantifiers as Giannakidou (2000) argues, then these must move outside the scope of the negative operator at or before LF in order to scope over the negation and derive the reading $\forall x, \neg P(x)$.

Zeijlstra (2004, 238–9) observes that universal quantifiers typically do not scope over negation. However, Giannakidou (2000) argues that only concordant negative universals scope over negation, and that this blocks other universals from doing so. Despite the counter-arguments presented by Zeijlstra (2004, 238–9) to the universal quantifier approach, only an analysis of strict negative concord in which pre-verbal concordant negative words are quantificational rather than indefinites accommodates patterns of non-inversion in strict negative concord at successive stages of the Jespersen Cycle.

Under an analysis that treats concordant negatives as universal quantifiers, the derivation of (190a) will proceed in two steps (193) and (194). (193) shows the derivation at the point when multiple agree takes place.

(193)

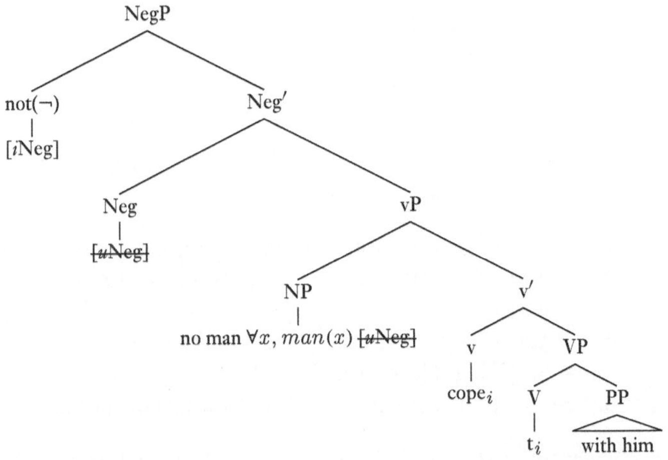


As it is a universal quantifier rather than an indefinite, the subject can move across the negative marker to spec,TP after its [*u*Neg] feature has been checked and deleted, in the manner of all subjects, to value its case features against the case feature of $T^0$ and value the $\phi$ features on $T^0$. (194) illustrates this stage of the derivation.

(194)

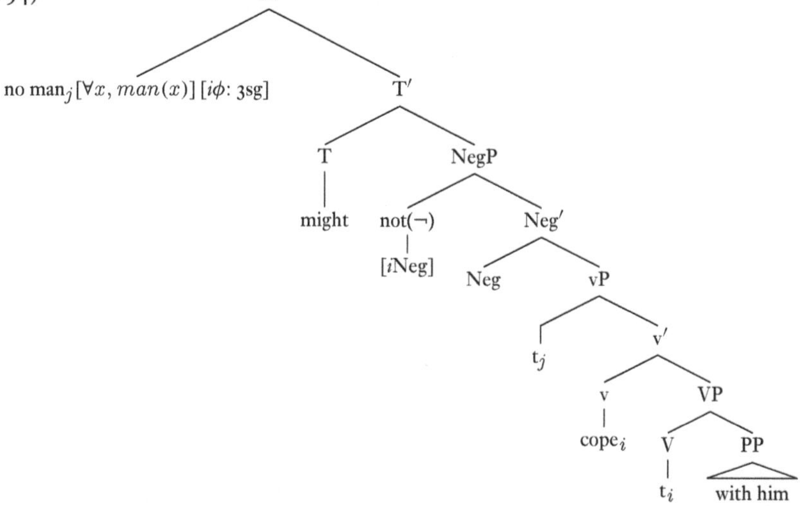


The analysis of concordant negative items as universal quantifiers makes predictions about the interpretation of pre-verbal negatives in strict negative concord. Although universals $\forall x, \neg P(x)$ and existentials $\neg\exists x, P(x)$ are logically equivalent, they differ in terms of presupposition and referentiality. $\forall x, \neg P(x)$ presupposes the existence of a set of individuals to which the predicate does not apply. On the other hand, in $\neg\exists x, P(x)$ the set to which the predicate applies is empty. As Giannakidou (2000) argues for Greek *kanenas* 'nobody' or *tipota* 'nothing'[14], concordant negative universal quantifiers are emphatic in some sense, referring to sets of entities that are discourse-given.

In (195), the use of a concordant negative universal quantifier introduces an existential presupposition that is the most appropriate reading in the context. Here it is not that Christ has no answer to the judge. Instead *nan þing* 'nothing' refers to a set of discourse-given or inferable answers that he could give to his accusers that he chooses not to give. At this point in the text, the reader knows the answer is that Christ is the Son of God.

(195) nan þing he ne answarode
no thing he NEG answered
'he did not give any answer'
(11th century; wsgosp,Mt_[WSCp]:27.12.2014)
$\forall x, thing(x) \wedge \neg answer(he, x)$

[14] I effectively make a similar claim that early English has two sets of concordant negatives, indefinites and universal quantifiers but that these are homophonous. Strict negative concord dialects have both existential and universal quantifier concordant negative items whereas non-strict negative concord dialects have only existentials, instead using a negative quantifier where that negative quantifier is outside the scope of the negative marker.

The derivations for (195) with *ne* [*i*Neg] and no negative inversion is given in (196) and (197). (196) shows the derivation at the point when multiple agree takes place. (197) shows the the complete clause structure.

(196)

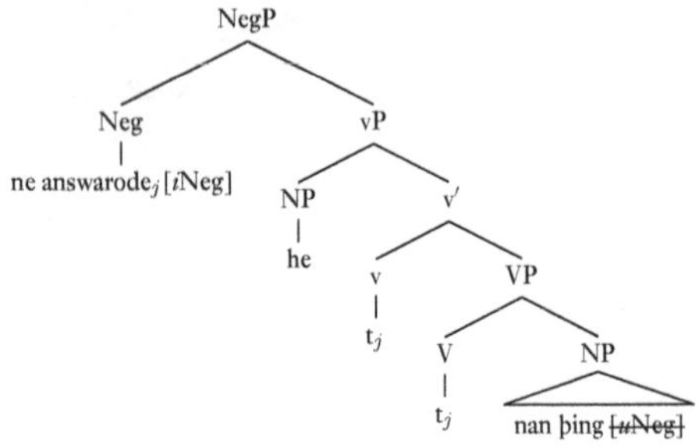


(197)

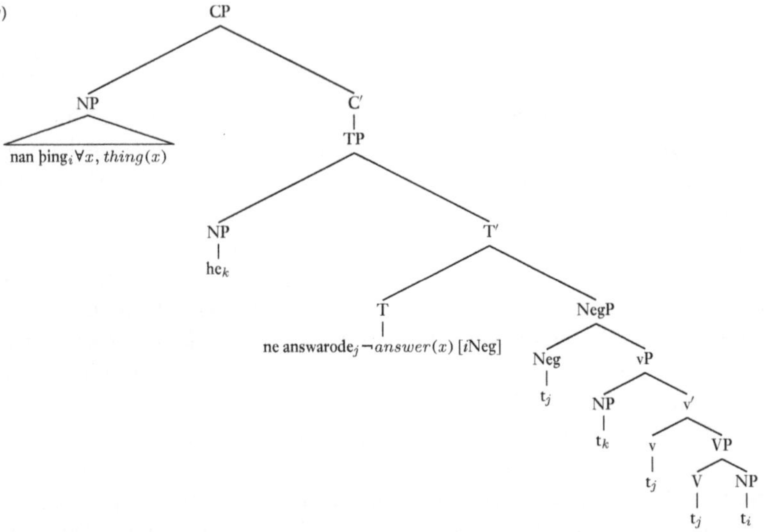


### *8.6.3 Negative Spread*

Finally, how does the derivation of (195) with negative doubling differ from the derivation of (198) with negative spread? The derivation of (198) with *ne* [*u*Neg] and negative inversion is given in (199) and (200). (199) shows the derivation at the point when multiple agree takes place. (200) shows the the complete clause structure.

(198) Nouther hwit ne blac ne nemmet he in his ordre
Neither white nor black NEG names he in his order
'He does not name in his order either white or black'
(13th century; ANCRIW,I.48.84)

(199)

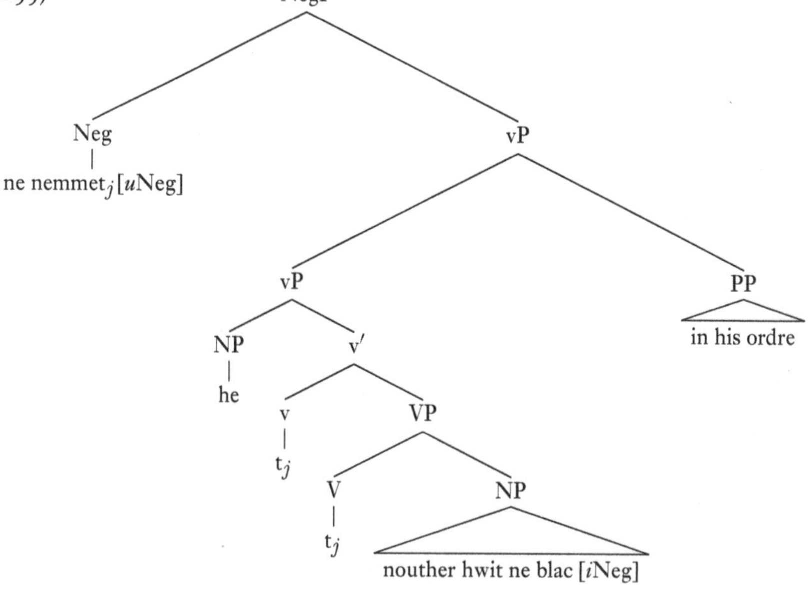


In (199) the configuration of [uNeg] and [iNeg] features is the wrong way around for Reverse Agree to operate. Therefore, the negative constituent [iNeg] must move either overtly before spellout or covertly after spellout for Reverse Agree to take place, as in (200). Under standard assumptions that negative operators must appear in a position of sentential scope at LF, when negative quantifiers have [iNeg] features and function as negative operators their movement out of VP to a higher scope position (spec,NegP or higher), either overtly or at LF, is generally assumed. In this case, movement of *nouther hwit ne blac* to spec,CP is focus movement, hence it is overt and co-occurs with T to C movement. We might derive focus movement to spec,CP as follows: $C^0$ has [iFocus], [uNeg*] which attracts *nouther hwit ne blac* [iNeg] to spec,CP before spellout. Following Roberts and Roussou (2003), the * diacritic marks a feature that must must be overtly realised at PF – one that is satisfied by morphologically overt internal or external merge. The * diacritic therefore forces internal Merge of the negative *nouther hwit ne blac* as the specifier of $C^0$. The category C* must also be overt at PF, and this requirement is satisfied by verb-movement to C.[15]

[15] There are a number of different approaches to head movement, discussed by Roberts (2010). In the present discussion, nothing very much follows from the precise analysis of V to C movement I adopt, so I leave issues of head movement to one side. What seems clear is that when focused, the head $C^0$ itself must be spelled out at PF by T to C movement.

(200)

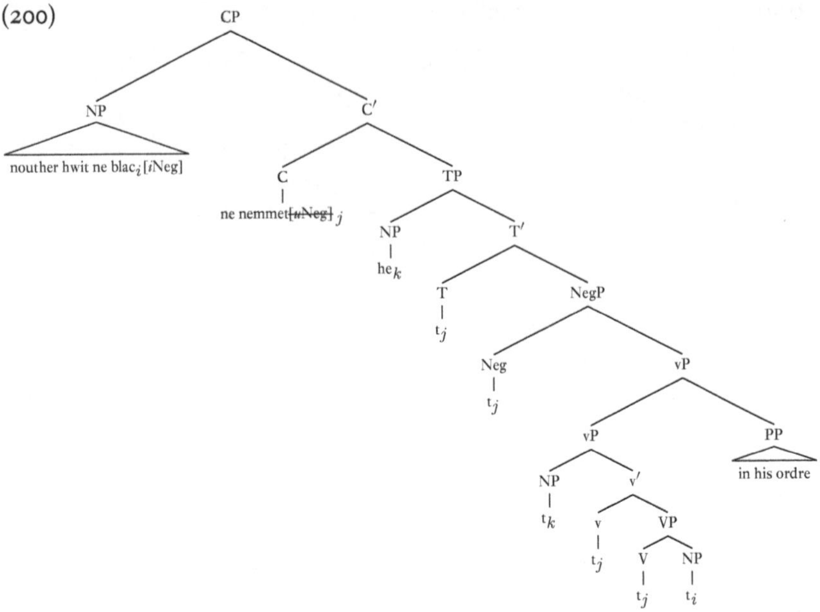


## 8.7 Summary

This chapter argued that changing patterns of negative inversion provide evidence for a quantifier cycle in early English proceeding in parallel with and interacting with the Jespersen Cycle. The negative inversion data are problematic to an analysis of negative concord where concordant negatives are indefinites. However, under an analysis of concordant negatives as quantifiers, we can maintain empirically well-founded links between negative inversion, negative operatorhood, the quantifier cycle and the Jespersen Cycle. The changes affecting *ne* in sentential negation contexts also affect *ne* in negative concord contexts. This interaction between weakening of *ne* and the quantifier cycle results in negative spread replacing negative doubling in many EME clauses, bleeding the Jespersen Cycle of negative doubling contexts in which *not* can be introduced. Chapter 9 shows how we can model the changing distribution of ME negative concord under this analysis, and relate it to the Jespersen Cycle.

# 9 The Loss of Negative Concord

## *Interaction Between the Quantifier Cycle and the Jespersen Cycle*

### 9.1 The Jespersen Cycle and Negative Concord in Middle English

This chapter examines patterns of variation and change in negative concord, arguing that new constraints on negative concord arise from the interaction between the Jespersen Cycle and the quantifier cycle as $ne_1$ and $ne_2$ are lost in Middle English. Negative concord with *not* is rare in ME. It is not generalised to all negative concord constructions. Instead, the frequency of negative concord declines during ME. Formal syntactic accounts of the relationship between the Jespersen Cycle and negative concord do not account for this. This chapter argues that changes in the availability and form of negative concord during ME arise not from formal syntactic generalisations linking the availability of negative concord to particular stages of the Jespersen Cycle, but instead through the interaction of the two processes of change – the Jespersen Cycle and the quantifier cycle – I described in Chapter 8. The patterns of variation that emerge out of this interaction between the Jespersen Cycle and the quantifier cycle and persist as Present-day English variation between no-negation (201a) and not-negation (201b).

(201) a. I said nothing (no-negation)
b. I didn't say anything (not-negation)

Table 9.1 presents the frequencies of clauses with negative arguments in the PPCME2. Four contexts are distinguished: three types of negative concord, with *ne* as in (202a), *ne...not* as in (202b) and *not* as in (202c); and clauses like (202d) without negative concord, in which the negative argument is the only negative word. The frequencies of *ne, ne...not* and *not* in sentential negation contexts (clauses in which *ne, ne...not* and *not* are the only negative words) are given alongside for comparison.

(202) a. he ne may noþing wel conne
he NEG may nothing well know
'He may not know anything well'
(14th century; AYENBI,117.2247)

Table 9.1. *A comparison of the frequency of* ne, ne...not *and* not *in negative doubling involving a negative argument, and sentential negation contexts.*

| | Negative argument | | | | | Negative markers only | | | |
|---|---|---|---|---|---|---|---|---|---|
| Period | *ne* | *ne...not* | *not* | no neg marker | Total | *ne* | *ne...not* | *not* | Total |
| 1150–1250 | 91.2% (n=394) | – (n=0) | – (n=0) | 8.8% (n=38) | 432 | 60.6% (n=436) | 38.5% (n=277) | 1.0% (n=7) | 720 |
| 1250–1350 | 90.0% (n=127) | 0.7% (n=1) | – (n=0) | 9.3% (n=13) | 141 | 22.9% (n=166) | 67.7% (n=490) | 9.4% (n=68) | 724 |
| 1350–1420 | 8.7% (n=77) | 0.8% (n=6) | 1.6% (n=14) | 89.1% (n=793) | 890 | 1.9% (n=43) | 10.5% (n=236) | 87.5% (n=1959) | 2238 |
| 1420–1500 | 0.3% (n=2) | – (n=0) | 2.0% (n=17) | 97.8% (n=849) | 868 | 0.7% (n=14) | 1.0% (n=18) | 98.3% (n=1842) | 1874 |
| 1500–1570 | 0.0% (n=0) | 0.0% (n=0) | 1.0% (n=12) | 99.0% (n=1202) | 1214 | 0.03% (n=1) | 0.0% (n=0) | 99.97% (n=3144) | 3145 |

b. Thou ne shalt nat eek make no lesynges in thy
You NEG shall not also make no falsehoods in your
confessioun...
confession...
'You shall also not tell any lies in your confession...'
(14th century; CMCTPARS,325.C1a.1581)

c. and he wolde not make noo confession unto no pryste...
and he would not make no confession to no priest...
'and he would not make any confession to any priest...'
(15th century; GREGOR,233.2474)

d. they hade no power to speke
they had no power to speak
'They had no power to speak'
(15th century; CMSIEGE,93.728)

Up to c.1420, negative concord typically involves the negative marker *ne* in examples like (202a). After 1350, there is a fairly steep decline in negative concord with *ne* which seems to parallel the loss of *ne* in sentential negation contexts[1], but this is not matched by a corresponding increase in negative concord with *not*. After 1350, we find some instances of negative doubling involving *ne...not* as in (202b), or *not* as in (202c), but these are not as frequent as the earlier pattern involving *ne*. Negative doubling with *(ne)..not* is extremely rare. It occurs in no more than 2.4% (n=20/890) of

[1] Late Middle English negative concord with *ne* is restricted to a particular South East Midlands and London dialect chiefly represented in the language of Chaucer (see also Jack (1978a)). This dialect also exhibits a higher frequency of bipartite *ne...not* than other dialects.

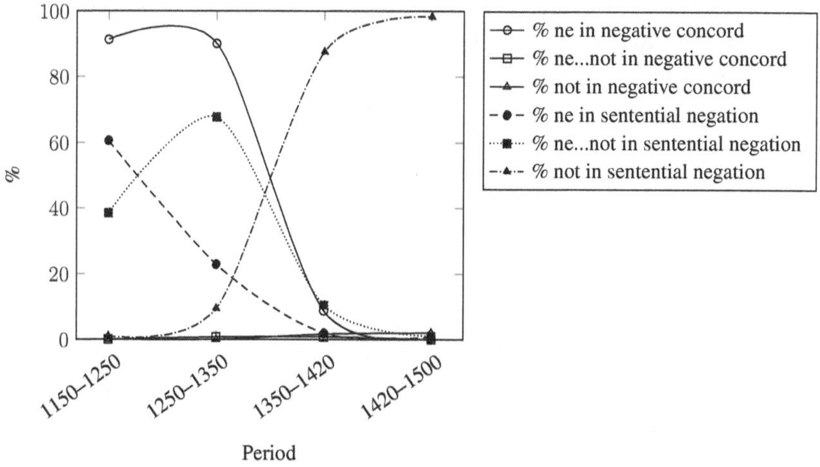


Figure 9.1 The frequencies of *ne* and *(ne)...not* in negative concord and sentential negation environments.

clauses in the period 1350–1420 and 2.0% (n=17/868) of clauses in the period 1420–1500.

Figure 9.1 illustrates that the main difference between sentential negation contexts and negative concord contexts is that while the loss of *ne* is dependent on the introduction of *not* in sentential negation contexts, the loss of *ne* is independent of the introduction of *not* in negative concord. In sentential negation contexts the loss of *ne* results in the introduction of *not*, whereas in negative concord contexts it does not.

## 9.2 Formalising the Relationship between the Jespersen Cycle and the Availability of Negative Concord

The early English pattern reported in Table 9.1, in which negative concord with *not* is infrequent, is but one instance of a pattern common across many languages. Jespersen observes a relationship between the form of the negative marker and the availability of negative concord, stating:

> There is one very important observation to be made, without which I do not think we will be able to understand the matter, namely that repeated negation [negative concord] becomes an habitual phenomenon in those languages only in which the ordinary negative element is comparatively small in phonetic bulk... If this repetition is rarer in modern English and German than it was formerly, one of the reasons probably is that the fuller negative *not* and *nicht* have taken the place of the smaller *ne* and *en*. (Jespersen, 1924, 333)

Negative concord tends to involve negative markers that are proclitic on the finite verb. Rowlett (1997, 1998) formalises this observation syntactically as 'Jespersen's Generalisation' (203).

(203) Jespersen's Generalisation (Rowlett, 1998, 87, ex.2): A language is an NC [negative concord] language iff the regular marker of pure sentential negation is not associated with spec,NegP.

In making this proposal, Rowlett restricts the availability of negative concord to stage one of the Jespersen Cycle (where the negative marker *ne* negates the clause on its own) and stage two of the Jespersen Cycle (where *ne* is the negative marker and the concordant adverbial *not* stands in negative concord with it). The negative marker *not* at stages three and four of the Jespersen Cycle is predicted to be incompatible with negative concord. However, we find negative concord with *ne...not* in the fourteenth and fifteenth centuries, as in (202b); and with *not* from the fifteenth to seventeenth centuries, in (202c). Rowlett's proposal is too restrictive to account for all patterns of early English negative concord.

Haegeman and Zanuttini (1996) relate the availability of negative concord to the Jespersen Cycle in a different way. They claim that the loss of negative concord correlates with loss of the negative head. Zeijlstra (2004) makes a similar generalisation, but within a Minimalist syntactic framework. He argues that negative concord relies on a multiple agree relation, which is initiated by an LF-uninterpretable negative feature on a negative head $Neg^0$ that acts as the probe. Under this approach, a phonologically null negative head is required to accommodate negative concord at stage four of the Jespersen Cycle, in examples like (202c). In LME, *not* occupies the specifier of a NegP which has a null head (see the discussion of NegP in Section 5.3.3). Allowing all languages with $Neg^0$ to be negative concord languages predicts that negative concord with *not* will be productive in LME, so it does not – on its own – explain why the frequency of negative doubling with *not* is so low.

Under the system of Reverse Agree I adopt in Chapter 8, it is not the negative head $Neg^0$ that is the probe in negative concord, but the concordant negative arguments and adverbials. Thus we cannot parameterise the availability of negative concord in terms of the presence or absence of NegP, its internal structure, or differences in the feature specification of $Neg^0$. Instead, the availability of negative concord follows from the feature specification of concordant negative words. When concordant negative words have uninterpretable negative features, at stage three of the quantifier cycle, they must occur within the scope of an [iNeg] feature present on another negative word, hence in negative concord with that negative word. Conversely, when each negative word is specified with an LF-interpretable [iNeg] formal feature, none of them will act as probes to establish Agree relations with other negative words, so negative concord will not occur. Variation and change in the feature specification of individual negative items provides a mechanism

Table 9.2. *The interaction of the quantifier cycle and the Jespersen Cycle.*

| | 8th–12th century *ne*$_1$ | 13th–15th century *ne*$_2$ | 16th century–PDE |
|---|---|---|---|
| **Quantifier Cycle:** | ne$_1$[iNeg]...negQ[uNeg] (negative doubling) | ne$_2$[uNeg]...negQ[iNeg] (negative spread) | negQ[iNeg] (no-negation) |
| **Jespersen Cycle:** | | ne$_1$[uNeg]...not[iNeg]...negQ[uNeg] (negative doubling) | not[iNeg]...any+N (not-negation) |

for variation and change in negative concord: loss of negative concord follows from the loss of negative items with [uNeg] features.[2]

This approach makes the cross-linguistic correlation between verbal-proclitic negative markers and negative concord observed by Jespersen (1924, 333) difficult to formalise. Reverse Agree makes changes to the features of negative markers (the Jespersen Cycle) and changes to the features of negative words (the quantifier cycle) formally independent of each other. Particular patterns of negative concord can only emerge out of the interaction of these two formally independent changes. Examples like (202a), (202b) and (202c) show that our analysis should not rule out negative concord at any stage of the Jespersen Cycle, so how can we explain the observation that the loss of negative concord correlates with the loss of *ne* in many languages and account for the low frequency of *not* in negative doubling?

The analysis schematised in Table 9.2 treats the relationship between the Jespersen Cycle and the quantifier cycle as a diachronic one rather than a formal one. Both the Jespersen Cycle and the quantifier cycle introduce new strategies to mark sentential scope for negation as *ne*$_1$ [iNeg] is lost. The quantifier cycle therefore bleeds the Jespersen Cycle in EME. The key to understanding the distribution of negative concord with *not* is that it is negative doubling, in which negation is marked at LF by the negative marker. *not* replaces *ne*$_1$ in negative doubling. However, the quantifier cycle eliminates negative doubling from the grammar, by turning negative doubling contexts into negative spread contexts in which *not* cannot appear. In negative spread, the negative quantifier is a negative operator and *ne*$_2$ is in concord with it. Therefore negative spread contexts have no need of the negative operator *not* to mark negation at LF. The result: negative spread with *ne*$_2$ and negative doubling with *not* are in complementary distribution. Even after the loss of *ne*$_1$, negative concord contexts retain *ne*. They exhibit negative spread between a negative quantifier and concordant *ne*$_2$, rather than negative

2 Zeijlstra (2008) argues that negative words in double negation languages are all specified with a semantic $\neg$ operator rather than a formal feature, hence they do not enter into feature-checking dependencies with other negative items. This achieves the same result, the loss of negative concord.

doubling between a negative marker (EME $ne_1$ or LME *not*) and a concordant negative word.

## 9.3 Modelling the Relationship between the Jespersen Cycle and Negative Concord

Frisch (1997, 33) explains that clauses with negative indefinites are problematic to a model of the Jespersen Cycle, because neither the loss of *ne* nor the introduction of *not* proceed in the same way, or even at the same rate, in sentential negation contexts and negative doubling contexts.[3] Therefore, he argues, sentential negation contexts and negative concord contexts cannot be incorporated into a single model. However, this is because Frisch's model does not take into account the distinction between $ne_1$ and $ne_2$, or the existence of the quantifier cycle and its interaction with the Jespersen Cycle.

### 9.3.1 *The Loss of Negative Concord with ne*

Negative concord with *ne* comprises two distinct types of negative concord at successive stages of the quantifier cycle – negative doubling with $ne_1$ at stage three of the quantifier cycle and negative spread with $ne_2$ at stage four of the quantifier cycle. The loss of negative concord with *ne* therefore proceeds in two steps. The replacement of negative doubling with $ne_1$ by negative spread with $ne_2$ is a reflex of the weakening of *ne*. Therefore it is concurrent with stage three of the Jespersen Cycle, from around the tenth to thirteenth centuries. The loss of negative spread with $ne_2$ is concurrent with the loss of $ne_2$ at stage four of the Jespersen Cycle from around the thirteenth to fifteenth centuries.[4]

Table 9.3 shows that from c.1250 onwards, the loss of *ne* follows the same course (both in terms of rate and intercept) whether it co-occurs with *not* or with a negative argument. This parallel distribution follows if LME negative concord with *ne* is negative spread with $ne_2$.[5] Variation between texts provides further evidence of the parallel between $ne_2$...*not* and negative concord with *ne*. As Table 9.4 shows, the frequencies of $ne_2$...*not* and negative concord with *ne* co-vary across texts in the period 1350–1420, but the frequencies of $ne_1$ and negative concord with *ne* do not.

3 Logistic regression models estimate that in PPCME2 sentential negation contexts the overall loss of *ne* proceeds at a rate of around 6 logits/century and in negative doubling contexts at a faster rate of nearly 8 logits/century.

4 Other forms of negative spread are lost later, during the sixteenth century (see Nevalainen (1996) for discussion of the mechanisms and factors involved).

5 The negative inversion data discussed in Chapter 8 indicate that, from the thirteenth century, most negative concord with *ne* is negative spread with $ne_2$.

Table 9.3. *The distribution of ne in clauses with negative arguments and clauses with not.*

| | Negative argument | | | *not* | | |
|---|---|---|---|---|---|---|
| *Period* | *ne* | Total | % *ne* | *ne* | Total | % *ne* |
| 1250–1350 | 128 | 141 | 90.8% | 490 | 558 | 87.8% |
| 1350–1420 | 83 | 793 | 10.5% | 236 | 2195 | 10.8% |
| 1420–1500 | 2 | 849 | 0.2% | 18 | 1860 | 1.0% |

Table 9.4. *The distributions of ne in negative concord and sentential negation contexts by text, period 1350–1420.*

| | Negative argument | | *not* | | $ne_1$ | |
|---|---|---|---|---|---|---|
| Text | % *ne* | n | % *ne* | n | % *ne* | n |
| Chaucer[6] | 37.9% | (n=47/124) | 39.7% | (n=81/204) | 1.9% | (n=4/208) |
| Others | 5.4% | (n=36/669) | 7.8% | (n=155/1991) | 0.1% | (n=2/1993) |

### 9.3.2 *The Introduction of Negative Doubling with not*

The quantifier cycle bleeds the introduction of negative doubling with *not*. One effect is that the rates at which *not* is introduced in negative concord and sentential negation contexts differ. It appears to spread at a slower rate in negative concord – hardly to spread at all in fact – because *not* can only compete with $ne_1$ in a subset of ME negative concord contexts – those involving negative doubling, not negative spread.[7]

The introduction of negative doubling with *not* does not go to completion, but remains variable. Constraints on negative doubling with *not* therefore result from the intersection of the Jespersen Cycle and the quantifier cycle. Their intersection is modelled by the formula in (204), based on Kroch (1989, 203, ex.1), where the first part represents the logistic function for the loss of negative doubling under the quantifier cycle, with intercept $k_1$ and slope $s$;

[6] The texts by Chaucer in the PPCME2 comprise samples from 'The Tale of Melibee' (CMCTMELI), 'The Parson's Tale' (CMCTPARS), his translation of 'Boethius' (CMBOETH), 'The Treatise on the Astrolabe' (CMASTRO) and 'Equatorie of the Planets' (CMEQUATO).

[7] If we could reliably separate negative doubling and negative spread contexts in ME, we would properly circumscribe contexts for the introduction of *not*. In negative doubling contexts, it should be introduced through competition with $ne_1$ proceeding at the same rate as in sentential negation contexts. However, it is difficult to determine which of the contexts involving negative concord with ME *ne* represent negative doubling with $ne_1$ and which negative spread with $ne_2$. This difficulty is reflected in the terminology I use in the remainder of the chapter – negative doubling with *not*, but negative concord with *ne* – since, with post-verbal negative words, it is difficult to distinguish whether this is negative doubling or negative spread. While Chapter 8 argued negative doubling and negative spread are distinct when a negative argument or adverbial appears clause-initially, unfortunately, such diagnostic contexts are too infrequent to permit detailed regression analysis.

and the second part represents the logistic function for the introduction of *not*, with intercept $k_2$ and slope $s$. $t$ is time and $e$ is a constant (Euler's number, 2.718).

(204) **Probability of negative doubling with *not*:**
P negative doubling $\left(\frac{e^{k_1-st}}{1+e^{k_1-st}}\right)$ x P *not* $\left(\frac{e^{k_2+st}}{1+e^{k_2+st}}\right)$

The slopes (rates of change) for both changes should be identical, as they are both reflexes of the loss of *ne*$_1$, although they are inverses of each other. The decrease in negative doubling is a negative slope hence $-s$, the increase in *not* is positive hence $s$. The change proceeds at a constant rate in all contexts, with contextual differences in the frequencies of negative doubling with *not* arising from variation in the two intercept parameters $k_1$ and $k_2$.

We do not have reliable independent estimates for these intercepts in negative doubling contexts, so it is not possible to model the frequencies of negative doubling with any precision using this formula. Were we able to estimate these intercept parameters, the interaction between the quantifier cycle and the Jespersen Cycle in (204) should predict the frequency of negative doubling with (*ne*)...*not*. Figure 9.2 shows it is quite possible to replicate the distributions of *not* in Figure 9.1 using the formula (204) and a pair of arbitrary intercepts $k_1$ and $k_2$. There are two lines representing the frequency of *not* in negative concord. The first (dashed line) results if the intercepts for the introduction of *not* (Jespersen Cycle) and loss of negative doubling (quantifier cycle) are the same. This derives an overall frequency of negative doubling with *not* that is too high. The second (solid line), which approximates more closely the distribution of *not* in Figure 9.1, illustrates what happens when the intercept for the quantifier cycle is higher (3 times higher, as represented by the light grey dotted curve) than the intercept for

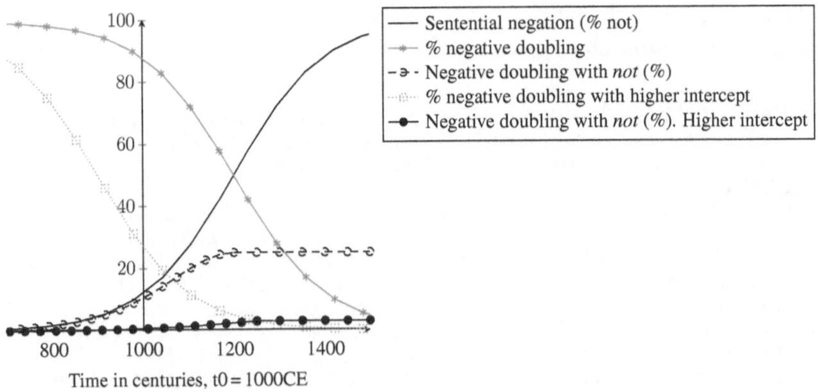


Figure 9.2 Negative doubling with *not* modelled using the formula in (204).

the Jespersen Cycle in sentential negation contexts. It therefore bleeds more contexts from the Jespersen Cycle.[8]

The constraints on negative doubling with *not* have two possible sources, corresponding to $k_1$, constraints on the loss of negative doubling; and $k_2$, constraints on the introduction of *not*. The constraints on *not* at stage three of the Jespersen Cycle that we observe in sentential negation contexts prior to its grammaticalisation there may also hold in negative doubling environments (hence these result from variation in $k_2$). For example, ME *not* is functionally marked in the same way in both sentential negation and negative doubling. However, the quantifier cycle may also bleed the negative doubling environments for *not* to different extents depending on variation in $k_1$ according to verb-type, clause type or other factors, with the result that negative doubling with *not* appears subject to constraints additional to those on *not* in sentential negation contexts.

Verb-type constraints are a good example. Table 9.5 shows that negative doubling with *not* is only found with lexical verbs or modals. Other verb types either exhibit negative concord with *ne* or lack negative concord.[9] We do not find these constraints on *not* in sentential negation contexts. Instead, they follow from the interaction of the quantifier cycle and the Jespersen Cycle – the quantifier cycle bleeds the introduction of negative doubling with *not* in all but a few clauses with lexical verbs.

### 9.3.3 *Functional Constraints on ME Negative Concord*

Nevalainen (1996, 267) observes that Late Middle English and Early Modern English negative concord becomes 'emphatic'. Functional constraints on negative doubling emerge as it begins to be lost from the fourteenth century onwards. Table 9.6 shows that negative concord is constrained by pragmatic activation in Late Middle English.[10] It subdivides clauses with negative arguments according to the position of the argument and the discourse-status of the proposition.

Discourse-old propositions are pragmatically activated – that is propositions that receive a prior mention in the discourse explicitly, or are that are implied within the discourse context, such that they are present to the audience's attention, as in (205).

[8] At least up to the midpoint of the curve, which represents the point at which the two changes intersect, and $ne_1$ is completely lost. At this point no-negation and not-negation enter stable variation and the curve levels out.

[9] Verb-type does not constrain LME negative concord with *ne* in the same way. Negative concord with *ne* is found with all verb types. This follows if negative concord with *ne* is negative spread with $ne_2$ rather than negative doubling.

[10] Table 9.6 only includes texts in which there is variation in the use of negative concord. Texts in which negative concord is uniformly absent are excluded from the figures.

Table 9.5. *The distribution of negative doubling in the period 1350–1420 according to discourse-function (D-new = discourse new, D-old = discourse old) and verb-type.*

| | modal+V | | lexical V | | HAVE | | copula BE | | existential BE | |
|---|---|---|---|---|---|---|---|---|---|---|
| Negative | D-new | D-old | D-new | D-old | D-new | D-old | D-new | D-old | D-new | D-old |
| *ne/n-* | 8.9% (n=9/101) | 17.5% (n=7/40) | 11.9% (n=10/84) | 25.0% (n=9/36) | 17.8% (n=11/62) | 33.3% (n=7/21) | 16.0% (n=8/50) | 36.3% (n=8/22) | 28.6% (n=14/49) | 41.6% (n=5/12) |
| *ne…not/not* | 5.0% (n=5/101) | 22.5% (n=9/40) | 3.6% (n=3/84) | – (n=0/36) | – (n=0/62) | – (n=0/21) | – (n=0/50) | – (n=0/22) | – (n=0/49) | – (n=0/12) |
| Subtotal neg-doubling | 13.9% (n=14/101) | 40.0% (n=16/40) | 15.5% (n=13/84) | 25.0% (n=9/36) | 17.8% (n=11/62) | 33.3% (n=7/21) | 16.0% (n=8/50) | 36.3% (n=8/22) | 28.6% (n=14/49) | 41.6% (n=5/12) |
| w/o neg-doubling | 86.1% (n=87/101) | 60.0% (n=24/40) | 84.5% (n=71/84) | 75.0% (n=27/36) | 82.3% (n=51/62) | 66.7% (n=14/21) | 84.0% (n=42/50) | 63.6% (n=14/22) | 71.4% (n=35/49) | 58.3% (n=7/12) |

Table 9.6. *The distribution of negative doubling in the period 1350–1420 according to discourse-function.*

| | Pre-verbal | | | | Post-verbal | | | |
|---|---|---|---|---|---|---|---|---|
| Negative | Discourse-new | | Discourse-old | | Discourse-new | | Discourse-old | |
| *ne*/n-proclitic | 6 | 5.6% | 10 | 19.2% | 48 | 17.3% | 33 | 33.3% |
| *ne...not* | 0 | – | 0 | – | 3 | 1.1% | 3 | 3.0% |
| *not* | 1 | 0.9% | 4 | 7.7% | 4 | 1.4% | 3 | 3.0% |
| Subtotal neg doubling | 7 | 6.6% | 14 | 26.9% | 55 | 19.9% | 39 | 39.4% |
| W/out neg doubling | 99 | 93.4% | 38 | 73.1% | 222 | 80.1% | 60 | 60.6% |
| Total | 106 | 100.0% | 52 | 100.0% | 277 | 100.0% | 99 | 100.0% |

(205) And in dede thou shalt love hym in swich wise that thou shalt
And in deed you ought love him in such ways that you ought
doon to hym in charitee as thou woldest that it were doon to thyn
do to him in charity as you would that it were done to your
owene persone. And therfore thou ne shalt don hym no damage
own person. And therefore you NEG ought do him no damage
in wikked word
in wicked words
'and indeed you ought love him in such a way that you ought do to him in charity as you would have done to yourself. Therefore you ought do him no damage through wicked words' (CMCTPARS,304.C2.660)

Such propositions favour negative concord, both negative concord with *ne* or negative doubling with *not*, significantly more than discourse-new propositions do.[11]

Pragmatic activation has a stronger effect on *not* than on *ne/n-*. In separate mixed-effects regression models for *ne* and *not*, each incorporating position and function as fixed effects and corpus text as a random effect, function of *ne/n-* has an effect size of 0.804, p=.005, while function of *not* has an effect size of 1.931, p<.001. This is key to understanding the relationship between the Jespersen Cycle and negative doubling. Chapter 6 showed that *ne* and *not* are functionally marked in Late Middle English. The functional constraints on negative doubling with *not* observed here are consistent with those on *not* in sentential negation contexts in the same period (*not* is favoured in discourse-old contexts, effect: 2.60, p<.001). Those on observed on negative

[11] Within a mixed-effects regression model incorporating function and position as predictors and corpus text as a random effect, function exhibits an effect size of 1.08, showing discourse-old propositions favour negative concord over discourse-new propositions. This effect is highly significant, p<.001.

concord with *ne* are similar to those on *ne*$_2$ in sentential negation contexts in the same period (discourse-old contexts favour *ne*$_2$, effect: 0.43, p=.17)[12], and quite distinct from those on *ne*$_1$ in earlier ME (1150–1250, effect: -3.23, p<.001; 1250–1350, effect: -2.64, p<.001). This is further evidence that LME negative concord with *ne* is negative spread involving *ne*$_2$, distinct from negative doubling involving either *ne*$_1$ or *not*.

Instead of a process of functional specialisation particular to negative concord contexts, the functional constraints on negative concord in late ME arise out of the interplay of more general functional constraints on *not* and *ne*$_2$. Discourse-functional constraints on the loss of *ne*$_2$, along with similar constraints on the introduction of *not* are independent of negative concord environments, but both cause negative concord to become specialised for contexts of pragmatic activation once *ne* begins to be lost.

## 9.4 Not-negation and no-negation

Interaction between the Jespersen Cycle and the quantifier cycle provides a principled explanation for the origins of PDE variation between not-negation and no-negation. ME negative doubling with *not* is the antecedent of not-negation, and negative spread with *ne*$_2$ the antecedent of no-negation. Negative doubling with *not* in (206) is replaced by *not*-negation, with non-veridical negative polarity items from the *any*-series standing in the scope of *not*.

(206) for he cowde not vndyrstond non Englysch.
'for he could not understand any English'
(CMKEMPE,81.1819)

Examples of not-negation like (207a) appear as early as c.1400. However, Nevalainen (1996) shows that the majority of concordant negative items are replaced by negative polarity items (NPIs) during the sixteenth century.

(207) a. And resoun of þis stondeth herynne þat God may not iuge
And reason of this stands herin that God may not judge
folily ony man;
erroneously any man;
'and the reason for this stands herin: that God may not judge any man erroneously'
(14th century, CMWYCSER,I,237.255. c.1400)

b. and as for the lawes of the realme, I truste I have not offendyd or brokyn anye of them.
'and as for the laws of the realm, I trust I have not offended or broken any of them.'
(16th century, MOWNTAYNE-E1-P1,181.71, c.1555)

12 This coefficient comes from Section 6.4.

Nevalainen's (1996) data demonstrate that the replacement of concordant negatives with NPIs is bound up with standardisation, connected with patterns of demographic change and social mobility in sixteenth-century London, and is especially prevalent amongst certain middle ranking professional social networks. The changes does not happen in all English varieties. Some non-standard varieties retain concordant negative words up to the Present-day (see for example Labov (1972), Smith (2001)).

Negative doubling with *not* is the antecedent of not-negation. Therefore, the null hypothesis is that Present-day English variation between no-negation and not-negation has its origins in the distribution of Middle English negative doubling with *not*, and that the constraints on its use are historically persistent, mirroring the constraints on ME negative doubling with *not*.

### 9.4.1 Not-negation vs. not-negation: Historical Persistence or Ongoing Change?

However, this runs counter to Tottie (1991a). She hypothesises that not-negation emerges in two stages. First she observes that some contexts for not-negation are ones in which no-negation cannot occur; or in which no-negation and not-negation are not semantically equivalent. As these contexts require a negative marker, *(ne)...not* simply replaces *ne* like it does elsewhere. While this seems empirically justified by the data Tottie (1991a) presents, it does not explain how not-negation enters into variation with no-negation.

She proposes that variation between no-negation and not-negation emerges during Middle English as *not* replaces *ne*. However, while *not* becomes the default marker of sentential negation by the fifteenth century, in clauses with indefinite elements it is not grammaticalised even in PDE. Instead it remains optional, and pragmatically marked as we saw in Section 6.6. To explain why the grammaticalisation of not-negation is incomplete, Tottie (1991a) hypothesises that the spread of not-negation remains an ongoing change in clauses with indefinites, even in PDE. It follows that not-negation and no-negation are in competition, rather than variation, throughout the Modern English period. If Tottie is right, the overall frequency of not-negation will show an increase over the course of the Modern English period. This is somewhat problematic: first, if *not* spreads through competition with $ne_1$, then how can we motivate its continued spread after $ne_1$ is lost; second, if the spread of not-negation is a reflex of the more general introduction of *not*, its spread should proceed at the same rate as *not* in other contexts and lead the change to completion before the Present-day.

### 9.4.2 Verb-type Constraints on not-negation

In a thorough analysis of data from the London-Lund Corpus, Tottie (1991b) identifies some grammatical constraints on not-negation. In particular, she observes that it is preferred with certain verb types, principally lexical verbs and AUX + lexical verb sequences, and dispreferred in negative existentials. Table 9.7 provides evidence for similar constraints on not-negation in both the written registers and the spoken conversation registers of the BNC.

Tottie (1991a) argues that this is an effect of lexical frequency. The most frequent verb types resist not-negation, even up to the present day. According to Tottie (1991a), not-negation is currently undergoing lexical diffusion. It occured first with the least frequent verbs, and is now gradually spreading to the the most frequent verbs such as BE and HAVE. She presents evidence for this from Helsinki Corpus data based on the frequency of not-negation with different verbs in the thirteenth to seventeenth centuries. However, if the verb-type constraints on not-negation are the result of lexical diffusion of *not*, it is odd that ME *not* does not spread in the same way at stage three of the Jespersen Cycle. The verb-type constraints on not-negation are not manifest on ME *not*, a fact for which Tottie's analysis does not account. However, in the model I propose, the verb type constraints can emerge out of the interaction of the Jespersen Cycle and the quantifier cycle. The quantifier cycle plays no role in sentential negation contexts, so the constraints it imposes can be particular to not-negation contexts.

Furthermore, Tottie does not examine the evidence for lexical diffusion within a a logistic model of change. Within a logistic model, a lexical diffusion based account implies that we should see (a) an increase in the overall frequency of not-negation over time, and (b) a gradual weakening of the verb-type constraints on not-negation as it spreads from most favoured to least favoured contexts. We know that negative doubling with *not* is subject to verb-type constraints even in the fourteenth century. Are these constraints the same as those on PDE not-negation, or do they change in later periods?

Table 9.7 compares verb constraints in written and spoken BNC data, seventeenth-century written data from the PPCEME and fourteenth century written data from the PPCME2. With HAVE (208) and BE (209) and (210), not-negation is particularly rare, but with lexical verbs (211) not-negation occurs more frequently, and with a modal (212) it is yet more frequent. The (a) example of each pair illustrates no-negation, the (b) example not-negation.

(208) a. He had no notion of it (BURNETROC-E3-P1,45.83)
b. Evil had not any Nature and Existence which might be properly so called (BOETHPR-E3-P2,153.305)

(209) a. Yett there is nothing under heaven which I desire more then to bee with thee. (HOXINDEN-1640-E3-P1,112.58)
b. There is not any of the batcholers in this country are incline-ing to marry this yeare that I heare of. (EOXINDEN-1660-E3-H,309.33)

(210) a. But this is no place to mention what thoughts I have had to and fro about these Matters (BOYLECOL-E3-P2,152.92)

Table 9.7. *The distribution of no-negation and not-negation by verb type in PPCME2 data from1350–1420, PPCEME data from 1640–1710 and BNC written and spoken data*

| | PPCME2, 1350–1420 | | | PPCEME, 1640–1710 | | | BNC written | | | BNC spoken[13] | | |
|---|---|---|---|---|---|---|---|---|---|---|---|---|
| Verb type | no-negation | not-negation | Total | no-negation | not-negation | Total | no-negation | not-negation | Total | no-negation | not-negation | Total |
| Existential BE | 100.0% (n=61) | 0.0% (n=0) | (n=61) | 97.9% (n=95) | 2.1% (n=2) | 97 | 99.0% (n=25256) | 1.0% (n=258) | 25514 | 97.1% (n=299) | 2.9% (n=9) | 308 |
| Copula BE | 100.0% (n=72) | 0.0% (n=0) | (n=72) | 95.8% (n=22) | 4.2% (n=1) | 23 | 97.9% (n=2813) | 2.1% (n=59) | 2872 | 88.0% (n=66) | 12.0% (n=9) | 75 |
| Lexical HAVE | 100.0% (n=83) | 0.0% (n=0) | (n=83) | 96.9% (n=62) | 3.1% (n=2) | 64 | 99.4% (n=16795) | 0.6% (n=102) | 16897 | 65.0% (n=39) | 35.0% (n=21) | 60 |
| Lexical verb (finite)[14] | 97.6% (n=123) | 2.4% (n=3) | (n=126) | 83.1% (n=69) | 16.9% (n=14) | 83 | 88.3% (n=7703) | 11.7% (n=1025) | 8728 | 19.9% (n=65) | 80.1% (n=261) | 326 |
| Aspectual AUX + lexical V | – | – | – | 63.6% (n=7) | 36.4% (n=4) | 11 | 78.7% (n=2266) | 21.3% (n=614) | 2880 | 7.5% (n=13) | 92.5% (n=160) | 173 |
| Modal + lexical V | 91.0% (n=127) | 9.0% (n=14) | (n=141) | 65.0% (n=39) | 35.0% (n=21) | 60 | 64.1% (n=1178) | 35.8% (n=657) | 1835 | 4.4% (n=7) | 95.6% (n=152) | 159 |
| TOTAL | 96.5% (466) | 3.5% (n=17) | 483 | 87.0% (n=294) | 13.0% (n=44) | 338 | 96.2% (n=56011) | 3.8% (n=2715) | 58726 | 44.4% (n=489) | 55.6% (n=612) | 1101 |

[13]The BNC spoken conversation data are a subset of the spoken data (s.conv) in the BNC. The sample comprises only the indefinite *-thing*, so *nothing* (no-negation) versus *not…anything* or *not…nothing* (not-negation).
[14]For the BNC written data, this group includes only inflected past tense forms, and 3rd person singular forms, in order to disambiguate finite and non-finite forms. Finite and non-finite forms are tagged separately in the PPCEME data so all are counted.

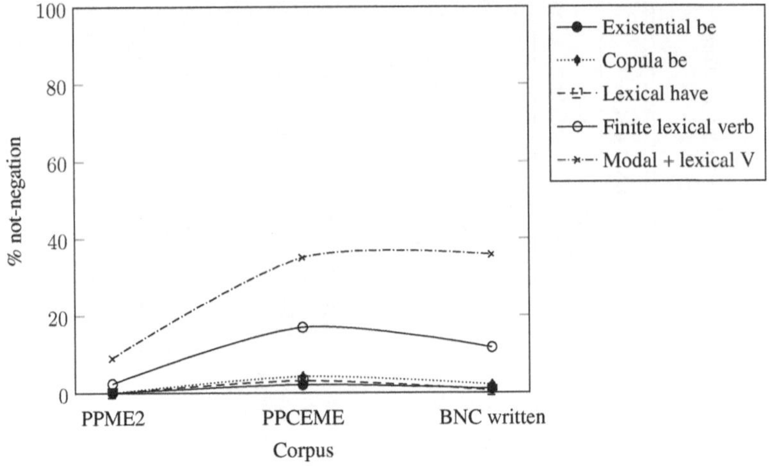


Figure 9.3 The frequency of not-negation in three written corpora.

b. I hope, my L=d=, it wase not any increase of y=r= indisposition w=ch= wase y=e= occasion I heard nothing from Kirby y=e= last post. 'I hope my Lady it was not any increase in your indisposition which was the occasion I heard nothing from Kirby by the last post' (CHATTON-E3-P1,2,150.32)

(211) a. But as for them, they believed none of those things; (TILLOTS-A-E3-H,418.21)

b. He did not know any thing of my coming there till I met him on the way. (LISLE-E3-P2,4.118.331)

(212) a. Monford being ignorant, could make him no direct answer (PENNY-E3-P2,142.23)

b. they can not comprehend any thing else (LOCKE-E3-P2,72.111)

Table 9.7 and Figure 9.3 show that in the BNC data, not-negation only occurs with any frequency when the predicate is a lexical verb, and is favoured in clauses with auxiliaries, particularly modals. The same patterns are observed in the Early Modern English (PPCEME) written sample dating from around three centuries earlier – the distinction between BE and HAVE on the one hand, and lexical verbs on the other are maintained, and appears robust given the size of the dataset. In fourteenth-century data from the PPCME2, we observe similar constraints on negative doubling with *not*. The overall frequency of *not* is somewhat lower in the ME data, perhaps because of ongoing competition with *ne*1. 17.6% (n=85/483) of the clauses retain negative concord with *ne*. However, the differences between the verb-types in later periods are entirely consistent with – and predictable from – the distribution of *not* in ME.

Table 9.8. *Logistic regression models of not-negation in PPCEME written (1640–1710) and BNC spoken (1985–1993) data.*

| | PPCEME data | | BNC spoken conv. data | |
|---|---|---|---|---|
| Context | Effect | p | Effect | p |
| Discourse-old | 2.36 | <.0001 | 2.41 | <.0001 |
| Lexical BE | -2.22 | .001 | -4.51 | <.0001 |
| Lexical HAVE | -1.43 | .076 | -1.71 | <.0001 |
| Modal AUX + lexical V | 0.92 | .041 | 0.97 | .003 |

Figure 9.3 shows that the frequencies of not-negation are consistent across the PPCEME and BNC written datasets and therefore do not provide any evidence of an overall increase in the use of not-negation during Modern English. These data suggest historical persistence of variation rather than ongoing change – specifically persistence of patterning which arises out of the interaction between the quantifier cycle and the Jespersen Cycle during Middle English, if not earlier. The frequencies of not-negation for each verb-type are remarkably similar in the BNC and PPCEME written samples. The frequency of not-negation is much higher in the spoken sample. However, as Biber (1995) observes, this may be a function of register rather than time.[15]

While the frequency of not-negation is much higher in the BNC spoken sample than in the either the Present-day (BNC) or Early Modern English written samples, the frequency of not-negation with BE and HAVE remains much lower than with lexical verbs. Table 9.8 presents two logistic regression models – one on the PPCEME written data from 1640–1710, the other on the BNC spoken conversation data. The models include verb type (reference level = lexical verbs) and discourse-function (reference level = discourse-new propositions) as fixed effects. Effect sizes are computed using treatment coding, with positive effects in contexts favouring not-negation over no-negation and negative effects in contexts favouring no-negation over not-negation.

Comparison of the written and spoken datasets in Table 9.7 provides no evidence that verb-type constraints are weaker in datasets where not-negation is more frequent overall nor do the verb type constraints on not-negation weaken over time. Instead, regression analyses show that both the verb type constraints and the discourse-functional constraints on not-negation are established before the seventeenth century and are simply historically

[15] This may be indicative of register variation along a spoken-written dimension rather than change. Biber (1995) distinguishes not-negation as a characteristic of spoken registers and no-negation as characteristic of written registers. In the data presented here, it is not possible to disentangle register variation from diachronic change, but this leaves open the interesting possibility that the register variation Biber (1995) observes represents the persistence of an earlier constraint on the introduction of Middle English *not* during the Jespersen Cycle. Unfortunately, the surviving ME data that comprise the PPCME2 are insufficiently varied according to register to test this hypothesis further.

persistent up to the Present-day. It is likely that both functional constraints (see Section 6.6) and verb-type constraints have their origins in the distribution of ME negative doubling with *not*, and arise out of the interaction between the quantifier cycle and the Jespersen Cycle, as $ne_1$ is lost in ME.

## 9.5 The Loss of Negative Concord: Interaction between Middle English Quantifier and Jespersen Cycles

The ME loss of negative concord resists an explanation that relates the availability of negative concord to particular stages of the Jespersen Cycle. The availability of negative concord is not restricted to particular stages of the Jespersen Cycle, though the Jespersen Cycle is very much implicated in its disappearance. Here, as elsewhere, a distinction between $ne_1$ and $ne_2$ in terms of LF-interpretability is fundamental. The loss of $ne_1$ at stage three of the Jespersen Cycle results in variation between negative doubling with $ne_2$...*not* and negative spread with $ne_2$. Where $ne_2$ [uNeg] enters into a feature checking dependency with the negative marker *not* [iNeg] we get negative doubling; where it enters into a dependency with a negative argument/adverbial [iNeg] we get negative spread. Variation between negative doubling and negative spread thus depends on the progress of the quantifier cycle in EME – to what extent negative arguments or adverbials are negative operators possessing an [iNeg] feature. Where negative arguments and adverbials have [iNeg] features, the introduction of *not* is blocked because the clause is already marked negative at LF. *not* can only appear in negative doubling contexts, but the quantifier cycle gradually eliminates these negative doubling contexts. In written English, negative doubling with *not* (and later, PDE not-negation) becomes the marked option.[16]

Negative doubling with *not* is the antecedent of PDE not-negation. Negative spread with $ne_2$ is lost as $ne_2$ is lost in the transition to stage four of the Jespersen Cycle, and is the antecedent of PDE no-negation. The verb-type and discourse-functional constraints on Present-day English variation between not-negation and no-negation arise out of the interaction between the Jespersen Cycle and the quantifier cycle in EME consequent on the loss of $ne_1$. This results in the persistence of stable variation between two functionally equivalent ways of marking sentential scope for negation in clauses with negative arguments or adverbials.

[16] Negative doubling – and later, not-negation – also become pragmatically marked because of the constraints on ME *not*'s introduction. Chapter 6 argued that in contexts where *not* is variable rather than categorical, its use remains pragmatically constrained.

# 10 Conclusion

Quantitative modelling of patterns of variation and change in corpus data provides an important means to identify syntactically, functionally and distributionally independent stages within the Jespersen Cycle, and provides an empirical basis to establish how formal change and functional change interact in corpus data. The findings of the study fall into two broad areas: characterisation of the grammatical and functional changes affecting early English negation, and their interaction; and the role quantitative models of diachronic change can play in our understanding of the Jespersen Cycle, grammaticalisation, and morphosyntactic change.

## 10.1 Grammar Competition Models

Most recent work on morphosyntactic change incorporates both qualitative and quantitative data. However, these quantitative analyses do not always take the form of thoroughly worked out statistical models of change over time, and as such do not allow us to realise fully the role of particular constraints or factors within a change. The accounts of functional change described in Chapter 6 provide a good demonstration of these issues.

Here, I develop the proposals made in Kroch (1989) to bring together a formal syntactic approach and a variationist approach. The grammar competition model becomes a diagnostic tool in exactly the way Kroch (1989, 235) concludes might be possible:

> Further work on historical change promises to extend the evidence linking patterns of change to grammars in competition, allowing us to understand changes better from the perspective of linguistic theory and eventually, perhaps, to refine grammatical analyses on the basis of the predictions they make about the patterning of usage in change.

Kroch (1989, 235, fn.29) elaborates:

> Once the principle that contexts change together when they are surface reflexes of a single grammatical competition becomes firmly

established, it may be possible, on occasion, to choose among grammars proposed on the basis of synchronic analysis by the predictions they make as to which contexts should change together.

Our syntactic framework must be sufficiently nuanced to allow theoretically plausible pathways of change via syntactic reanalysis, and to reproduce patterns of change observed at an empirical level. Those syntactic reanalyses must be learnable during language acquisition. By positing different elements within the analysis and specifying them in different ways, each formal analysis of change structures competition between forms differently, in terms of which forms are mutually exclusive, and hence what competes with what. Each analysis therefore makes a particular set of predictions about the distribution of competing forms at successive points within a change, and about how that change will progress over time. We can then test these predictions against observations of change in progress in diachronic corpus data; and where those predictions and observations are quantitative in nature, even make a statistical estimate of the extent to which the predicted and observed distributions of forms correlate. We can also evaluate the way each analysis structures the competition over both time and the contexts in which it occurs. Properly structured, change should proceed at a constant rate in each context where it occurs. This is simply an independence assumption – that the change in use is independent of the context of use. This is the null hypothesis, so any interactions between the change and the context in which it occurs must be plausibly motivated.

I am not the only one to demonstrate these points in practice and to utilise Kroch's model to inform a formal analysis of morphosyntactic change (see Pintzuk (1999), Han (2001), Han and Kroch (2000) and Warner (2005) for example). However, this is exactly what the present analysis of the Jespersen Cycle has achieved, particularly in using their distributional independence within changes in progress as a basis to distinguish two forms of *ne*. From a methodological standpoint, having demonstrated the value of this approach in tackling one of the best known case studies of grammaticalisation – the Jespersen Cycle – such an approach should prove insightful in disentangling other complex changes in which there are intersecting and interacting processes of change – for example, ascertaining how the cyclic changes in relativisation or modal marking strategies described by van Gelderen (2011) are structured in early English.

## 10.2 A Feature-based Analysis of the Jespersen Cycle

A Minimalist feature-based account allows us to distinguish changes to $Neg^0$ at the two interface levels LF (changes in its semantics) and PF (changes in its morphological form). Change at LF precedes change at PF. Modelling the changing distribution of *ne* in corpus data provides evidence to distinguish two forms of *ne* at successive stages of the Jespersen Cycle, each with different combinations of PF and LF interface properties, and each distributionally independent.[1] Several changes in the distribution of *ne* fall straightforwardly out of a model in which these two types of *ne* compete.

The basis on which to distinguish $ne_1$ and $ne_2$ is a difference in their ability to mark sentential scope negation at LF. The Jespersen and quantifier cycles comprise a series of intersecting reanalyses in the scope properties of individual negative items – whether they mark sentential scope negation themselves (as negative operators) or are licensed within the scope of another negative item (as anti-veridical NPIs). Each word that is negative in form is either interpreted as a negative operator ¬ at LF, or it is not. This can be formalised syntactically as a minimal difference in a single morphosyntactic (formal) [$\alpha$Neg] feature, as LF-interpretable [iNeg] or LF-uninterpretable [uNeg].[2] An interpretable neg-feature is one that identifies a word negative in both form and as contributing ¬ at LF. An uninterpretable neg-feature is one that identifies a word negative in form only. It schematises the Jespersen Cycle as in (213).

| | | | | | |
|---|---|---|---|---|---|
| (213) | Stage One (OE): | $Neg^0$ $ne_1$ | [*i*Neg] | | |
| | Stage Two (OE): | $Neg^0$ $ne_1$ | [*i*Neg] | + adverb $not_1$ | [*u*Neg] |
| | Stage Three (ME): | $Neg^0$ $ne_2$ | [*u*Neg] | + spec,NegP $not_2$ | [*i*Neg] |
| | Stage Four (14th c. on): | $Neg^0$ Ø | [*u*Neg] | + spec,NegP $not_2$ | [*i*Neg] |

$ne_1$ and $ne_2$ are involved in different processes of change, changes which pattern differently in corpus data. The patterning of competition between *ne, ne...not* and *not* within logistic regression models of change in progress shows that changes involving $ne_1$ and $ne_2$ stand in a particular relationship, as summarised in (214). There are three $Neg^0$ in two processes of competition: first, there is competition between LF-interpretable $Neg^0$ [iNeg] and LF-uninterpretable $Neg^0$ [uNeg], then competition between overt and null realisations of $Neg^0$ [uNeg]. Only by analysing the changes in this way does each proceed at a constant rate during Middle English. (214) dissociates change in the feature specification of $Neg^0$ and change in its morphological form. Although they follow distinct trajectories, the two changes are not independent – the patterning of change in corpus data shows that change in the feature-specifications of $Neg^0$ feeds change in its morphological form.

[1] Unlike Roberts and Roussou (2003, 28), I argue that interpretation of a feature at both LF and PF interfaces needs to be parameterised, and that variation and change in LF and PF interpretability are dissociated. There is empirical evidence for this: changes in interpretability at LF and PF proceed at different rates and subject to different constraints.

[2] Breitbarth (2009) makes a distinction between two forms of *ne*, but her characterisation of $ne_2$ as a non-veridical negative polarity item overgeneralises its distribution, contrary to the Middle English data. In Middle English, $ne_2$ has the properties of an anti-veridical negative polarity item, appearing only in concord with a syntactically local negative marker.

| | Morphosyntactic weakening of $Neg^0$ | | Morphological weakening of $Neg^0$ |
|---|---|---|---|
| | $Neg^0$ $ne_1$ [iNeg] + not [uNeg] | | |
| (214) | ↓ | ↗ | $Neg^0$ $ne_2$[uNeg] + not [iNeg] |
| | $Neg^0$ [uNeg] + not [iNeg] | | ↓ |
| | | ↘ | $Neg^0$ Ø[uNeg] + not [iNeg] |

Crucially, competition can occur at the level of individual morphosyntactic features, and change in the feature specification of $Neg^0$ may precede change in its morphological realisation, although equally the two changes may appear simultaneous. Competition between morphologically overt and morphologically null $Neg^0$ is restricted to $Neg^0$ [uNeg], because only negative items that are semantically uninterpretable can be morphologically null. Negatives that contribute ¬ at LF must be morphologically overt and identifiable. This ensures that the two changes stand in a feeding relationship as in (214). Structuring the Jespersen Cycle as in (214) accommodates the distributionally distinct forms $ne_1$ and $ne_2$ in two processes of change, each proceeding at a different but constant rate. In analyses that assume a single form of *ne* the loss of *ne* does not proceed at a constant rate.

Why does competition between two forms of *ne* result in the introduction of *not*? As (214) shows, change in the interpretability of the [Neg] feature at stage three of the Jespersen Cycle has different structural and functional reflexes. This is the key to understanding why $ne_2$ enters into a dependency with *not*. Within a Minimalist framework, structural and functional equivalence hold at different levels. Structural equivalence holds at the category level – the two $Neg^0$ [iNeg] and [uNeg] are mutually exclusive because they are both exponents of the same category, occupying the same syntactic position.[3] However, functional equivalence holds at the feature level – the [iNeg] feature can only occur once in a negative clause as it functions to mark negation at LF, thus all elements with [iNeg] features are mutually exclusive competitors for the function of marking negation at LF, irrespective of their category.

$Neg^0$ [iNeg] and $Neg^0$ [uNeg] are structural competitors – both negative heads ($Neg^0$). However, they are not functional competitors. Instead they are functionally distinct: $ne_1$ is a negative operator, contributing ¬ at LF; $ne_2$ is a concordant negative item, a non-veridical NPI. Conversely, $ne_1$ ($Neg^0$) and *not* (spec,NegP) are not structural competitors. Instead, they are mutually exclusive because they are functional competitors, both [iNeg], marking ¬ at LF. So loss of $ne_1$ is a single change with two reflexes, one structural, replacement of $Neg^0$ [iNeg] by its structural competitor $Neg^0$

3 Similarly, at a later stage, competition between *ne...not* and *not* emerges from competition between structural competitors for $Neg^0$ [uNeg], one morphologically overt, the other null.

[uNeg]; the other functional, replacement of $Neg^0$ [iNeg] by its functional competitor spec,NegP *not* [iNeg]. These two reflexes are tied together over the course of the change. They proceed at the same rate and subject to the same constraints. Without the kind of fine-grained sub-categorial/featural distinctions between lexical items that I make here, this link between changes to *ne* and the introduction of *not* is difficult to explain.

However, that different structural and functional reflexes arise out of a single parametric change is formalised under a Minimalist account. The dissociation between structural and functional competitors is resolved by innovation of a syntactic agreement dependency at stage three of the cycle. Distinguishing two forms of *ne* which stand in a different relationship to *not* explains why *not* does not simply replace *ne*. *not* gradually replaces its functional competitor $ne_1$ [iNeg]. However, *not* and $ne_2$ differ in both the interpretability of their neg-feature and their syntactic category. Therefore they are not competitors, and *not* does not replace $ne_2$. Instead *not* licenses $ne_2$ in it scope, so $ne_2$ and *not* co-occur. As $ne_2$ has to co-occur with *not* in order to be licensed, so the loss of $ne_2$ can only occur in clauses that already have *not*.

The changes in (214) provide a syntactically principled account for the weakening of *ne*[4] as a syntactic reanalysis [iNeg]>[uNeg][5], followed by a protracted period of competition between two featurally distinct competitors for the $Neg^0$ position, $Neg^0$ [iNeg] and $Neg^0$ [uNeg]. The patterning of both grammatical and functional constraints on $Neg^0$ [iNeg] and $Neg^0$ [uNeg] in diachronic corpus data, modelled using mixed-effects regression, supports this scenario. Competition between $Neg^0$ [iNeg] and $Neg^0$ [uNeg] proceeds at a constant rate in all environments, as does the spread of *not*, which – as explained above – is a reflex of this change.

This has implications for functionally-based accounts of the Jespersen Cycle. That *not* spreads at a constant rate in all environments shows functional extension or pragmatic unmarking plays no role in its spread. Pragmatic unmarking would manifest itself as different rates of change in pragmatically marked and unmarked environments, or as reweighting of the constraints on $ne_2$...*not* as it becomes more frequent. However, the role of pragmatic factors in conditioning competition between $ne_1$ and $ne_2$...*not* emerges as both significant and constant over time. $ne_2$...*not* spreads in

4 The reanalysis of *ne* and the grammaticalisation of *not* are interlinked. It is not clear which of these changes is a cause and which is a consequence, or if indeed, it is sensible to make such a distinction. Indeed, it is difficult to see how cause and effect could be distinguished empirically within a model of change when both are simultaneous. Even though the Jespersen Cycle is but one of many manifestations of the change affecting *ne*, it is not clear that this privileges the reanalysis of *ne* as the more primitive change.

5 The data described by Breitbarth (2009) and discussed in Chapter 5, in which *ne* behaves more like a non-veridical than an anti-veridical polarity item in some languages, such as French, suggests that *ne* may undergo a subsequent reanalysis in some languages, which should yield a third distributionally distinct form of *ne*.

competition with *ne*$_1$ at a constant rate in all pragmatic functions. It becomes pragmatically unmarked only when its competitor *ne*$_1$ is finally lost. There is no additional process of pragmatic unmarking operating on *not* at the constraint level.

## 10.3 Consequences of this Analysis

In the spirit of parametric analyses the analysis is (a) conceptually simple; (b) minimal, in that the difference between *ne*$_1$ and *ne*$_2$ reduces to the semantic interpretability of a single morphosyntactic feature - a difference which is one of the primitives of the Minimalist framework, and which should therefore be easily learnable; and (c) links several empirical observations (in this case changes) as reflexes of a single parameter (cf. Bobaljik and Thrainsson (1998) for example). The Jespersen Cycle is but one manifestation of the change in scope properties of *ne*. Concurrent changes to patterns of redundant negation, negative inversion and negative concord all provide evidence of the distributional independence of two forms of *ne* in competition. In each case, distinct patterns of redundant negation, negative inversion and negative concord map onto an underlying change in the scope properties of *ne* which can be formalised in terms of competition between *ne*$_1$ [iNeg] and *ne*$_2$ [uNeg]. This provides a formal syntactic link between several changes that occur in EME, and cluster at stage three of the Jespersen Cycle.[6]

Competition between *ne*$_1$ and *ne*$_2$ manifests itself in redundant negation environments. Redundant negation becomes restricted to contexts of wide scope negative concord as *ne*$_2$ [uNeg] replaces *ne*$_1$ [iNeg]. Thus we can distinguish between environments where negative markers are used redundantly, in which a negative marker is not interpreted truth conditionally; and environments where concordant negative items can appear in wide scope negative concord. The distribution of concordant *ne*$_2$ is subject to syntactic locality constraints, providing evidence that negative concord reduces to feature checking under Reverse Agree.

Negative inversion provides an important diagnostic for the scope properties of negative items. When they stand in clause-initial position, those which

[6] The analysis makes a further prediction – as Kroch (1989) observed for the loss of V to T movement and the rise of do-support – that all reflexes of a single parameter should change at the same rate, as they are in effect different manifestations of the same underlying change. It is not always practical to test this prediction in corpus data – particularly in the cases of redundant negation (Chapter 3), negative inversion (Chapter 4) and changes to negative concord (Chapter 8), relevant data are not always sufficiently numerous to model competition in sufficient detail to produce reliable estimates of rates of change. Furthermore, estimating rates of change implies consistency in the corpus data from one period to the next, with the data balanced across periods for other variables such as register/genre or dialect. Given the paucity of Middle English textual data, the PPCME2 deviates quite substantially from this ideal corpus. Larger corpora from later periods of English, such as the British National Corpus, alleviate these issues and provide sufficient resolution for more detailed analysis.

mark ¬ at LF trigger inversion and those that are negative in form only do not. Negative inversion is productive throughout the history of English, but the triggers for inversion change over the course of Old, Middle and Early Modern English. Changing patterns of negative inversion follow from changes in the scope properties of individual negative items. The loss of *ne*-initial clauses therefore follows from competition between *ne*$_1$ [iNeg] and *ne*$_2$ [uNeg] in EME. When negative inversion follows *ne*, it must be *ne*$_1$ [iNeg]. When *not* co-occurs with negative inversion, therefore, it is not a negative operator.

In OE, *nawiht* is a negative polarity item, and is compatible with inversion. Where it is an adverbial, it functions as a minimiser denoting minimal quantity on an extent scale associated with the predicate. Hence it only occurs with predicates that are gradable according to extent. The association between OE *nawiht* and psych-verbs follows from the fact that many of them are gradable by adverbials of minimal extent. Identifying weakening in these associations between *not* and particular predicates provides an argument to date its reanalysis to the eleventh or twelfth centuries. While in inverted clauses, *not* becomes a focus adverb marking contrastive polarity focus for a short time during late OE or EME, subsequent changes in the distribution, syntax and functions of *not* indicate its grammaticalisation as a negative operator during the twelfth and thirteenth centuries. This grammaticalisation of *not* is concomitant with the loss of negative inversion in clauses with *ne...not*. The association between grammaticalisation of *not* and loss of negative inversion with *ne* follows from reanalysis of *ne*$_1$[iNeg] ...*not*[uNeg] as *ne*$_2$[uNeg]...*not*[iNeg] during the twelfth and thirteenth centuries.

## 10.4 A Feature-based Analysis of the Quantifier Cycle

The negative inversion diagnostic also provides evidence for a quantifier cycle – clause initial negative arguments and adverbials come to trigger inversion more frequently during the course of Old and Middle English, indicating a change in their scope properties. They become negative operators during late OE and EME. That they do so while co-occurring with *ne* is evidence to distinguish two patterns of negative concord with *ne* – negative doubling and negative spread – in late OE and early ME, as schematised in (215) and (216).

(215) **Quantifier cycle stage three: negative doubling**
Negative argument/adverbial ~~[uNeg]~~ – Subject pronoun – *ne*$_1$ [iNeg] + V ...
(negation marked at LF by *ne*, negative argument/adverbial has [uNeg] features checked and deleted before movement to clause-initial position, no inversion)

(216) **Quantifier cycle stage four: negative spread**
Negative argument/adverbial[iNeg] – $ne_2$ ~~[uNeg]~~ + V – Subject pronoun …
(negation marked at LF by initial negative argument/adverbial, $ne_2$ has [uNeg] features checked and deleted, subject–verb inversion)

The quantifier cycle provides further evidence to distinguish two forms of *ne*, one a negative operator, the other a concordant negative item. Although competition between them manifests itself as different patterns of change in these two environments, the same two forms of *ne* compete in both sentential negation and negative concord during EME. In negative concord, the negative operator $ne_1$ marks negation in negative doubling; the concordant negative $ne_2$ stands in the scope of a negative argument or adverbial in negative spread. Analysing negative concord in terms of feature-checking via Reverse Agree allows us to to parameterise both the quantifier cycle and the Jespersen Cycle in exactly the same syntactic terms. This explains why they operate concurrently – both involve the loss of $ne_1$ in EME.[7] The key benefit of analysing the Jespersen Cycle and the quantifier cycle in the same syntactic terms is that this enables us to derive changes to the form and availability of negative concord as consequences of the way the quantifier cycle and the Jespersen Cycle interact over time.

The distribution of ME negative concord is problematic to accounts that attempt to link the availability of negative concord with particular stages of the Jespersen Cycle. Negative concord occurs at each stage of the Jespersen Cycle, but with varying degrees of productivity. Formal accounts of the relationship between negative concord and the Jespersen Cycle are not sufficient to explain either the coexistence of negative concord with *ne* and with *not* in LME, or why the productivity of LME negative concord with *ne* is so much higher than LME negative concord with *not*, despite the incipient grammaticalisation of *not* elsewhere. These distributional facts arise out of the interaction of quantifier and Jespersen Cycles. Their interaction predicts the restricted distribution of *not* in negative concord, and also the LME loss of negative concord (negative spread) involving $ne_2$.

By the fourteenth century, at stage three of the Jespersen Cycle, negative doubling with *not* (217) and negative spread (218) are mutually exclusive ways to mark sentential scope in negative concord, and are in complementary distribution at successive stages of the quantifier cycle. Negative doubling with *not* is the conservative variant. It is introduced when $ne_1$ is lost, only

7 This analysis requires us to postulate that concordant negative arguments and adverbials are quantificational rather than indefinites, contra Zeijlstra (2004) for example, but the relationship between negative inversion, scope and patterns of negative concord at successive stages of the Jespersen Cycle is otherwise difficult to explain. The syntax (object movement) and meaning (existential presupposition) of these concordant negative arguments/adverbials provides some independent evidence to corroborate this analysis.

where the quantifier cycle has not yet operated to reanalyse the negative argument or adverbial as a negative operator.

(217) **LME Quantifier cycle stage three: negative doubling**
Negative argument/adverbial ~~[uNeg]~~ – Subject pronoun – $ne_2$ ~~[uNeg]~~ + V … *not* [iNeg]

(218) **LME Quantifier cycle stage four: negative spread**
Negative argument/adverbial[iNeg] – $ne_2$ ~~[uNeg]~~ + V – Subject pronoun …

So the Jespersen Cycle proceeds in negative doubling environments just as in sentential negation environments, but the quantifier cycle progressively bleeds negative doubling contexts over the course of ME, turning them into negative spread contexts, in which *not* cannot appear. This is why negative concord appears to retain *ne* into LME – in most of these contexts *ne* is $ne_2$ in negative spread with an argument or adverbial that is itself the negative operator. It is telling that the loss of negative concord with *ne* patterns with the loss of $ne_2$ in clauses with *not*. Negative doubling in (217) is the antecedent of PDE not-negation. Negative spread in (218) is the antecedent of no-negation. Consequently, the interaction of the Jespersen Cycle and the quantifier cycle in EME establishes constraints on *not* which persist unchanged for a period of more than 500 years, explaining the continued variation between not-negation and no-negation in Present-day English and the constraints upon this variation in Present-day written and spoken corpus data.

Cross-linguistic work comparing the diachronies of the quantifier cycle and the Jespersen Cycle in a variety of languages may provide further detail and empirical support for this hypothesis. A more dynamic approach based on interacting processes of diachronic change may explain patterns of variation and change in negative concord that seem somewhat resistant to simple synchronic syntactic or typological generalisations.

## 10.5 Functional Change

The approach to functional change I propose allows us to test whether the particular role of pragmatic factors, and their relationship to the English Jespersen Cycle, is cross-linguistically generalisable. Do other Jespersen Cycles involve the same pragmatic or functional factors, and is their role in the cycle the same? Are Hansen (2009) and Hansen and Visconti (2009) in fact correct for French and Italian or not? If they show a common pattern, quantitative analyses of data from a greater number of Jespersen Cycles may be sufficient to settle the question of what role pragmatic change plays in the Jespersen Cycle.

In common with many other analyses of grammaticalisation phenomena, we may conclude that the grammaticalisation of *not* is fundamentally a process of functional extension or functional generalisation, in the sense that the functions of *not* move from being highly circumscribed by discourse factors to being primarily grammatical or semantic (the function of marking negation, irrespective of discourse context). However, these shifts in function – from minimiser to marker of discourse-givenness to pragmatically unmarked – do not involve gradual attrition of pragmatic constraints on the distribution of *not*. Instead, when we estimate the effects of pragmatic constraints on competing forms, we find long periods of stability in those functional constraints. This demonstrates that they are independent of the change, rather than arising during the change, or motivating it. Functional change results from changes in the way the competition between forms is structured. Pragmatic unmarking of *not* only takes place at the point when all the negation strategies that compete with *not* are lost. In a sense, then, pragmatic unmarking is the outcome of grammatical competition rather than a factor shaping the progress of that competition over time. Indeed, in clauses where more than one negation strategy remains possible – clauses with indefinites which may involve not-negation or no-negation – *not* remains pragmatically marked. The Middle English pragmatic activation constraints on *not* persist up to the Present-day. Of course, if *not* remains pragmatically marked until it becomes the grammatical default, the question of what drives its ME spread remains unresolved. While the present study has focused on the structure of the changes involved in the Jespersen Cycle - on both grammatical and functional levels – it has little to say on this point. S-curve shaped linguistic changes are so ubiquitous that this is a more general issue, whose answers must lie in general cognitive or psycholinguistic mechanisms, or processes of language acquisition. The Jespersen Cycle is not a special case, but – given a Minimalist syntactic analysis – it fits within the grammar competition model.

# References

Adger, D. (2006). Combinatorial variability. *Journal of Linguistics 42*, 503–530.

Adger, D. and P. Svenonius (2010). Features in minimalist syntax. In C. Boeckx (Ed.), *The Oxford Handbook of Syntactic Minimalism*, pp. 27–51. Oxford: Oxford University Press.

Adger, D. and G. Trousdale (2007). Variation in English syntax: theoretical implications. *English Language and Linguistics 11*, 261–278.

Adger, D. and G. Tsoulas (2004). Circumstantial adverbials and aspect. In S. Engelberg, J. Austin, and G. Rauh (Eds.), *The Syntax of Adverbials*. Amsterdam: John Benjamins.

Baayen, R. (2008). *Analyzing Linguistic Data: A Practical Introduction to Statistics using R*. Cambridge: Cambridge University Press.

Baayen, R., D. Davidson, and D. Bates (2008). Mixed-effects modeling with crossed random effects for subjects and items. *Journal of Memory and Language 59*, 390–412.

Baghdikian, S. (1979). *Ne* in ME and EModE. *English Studies 60*(6), 673–679.

Bates, D. M. (2010). *lme4: Mixed-effects Modeling with R*. http://lme4.o.r-forge.r-project.org/lMMwR/lrgprt.pdf

Bayer, J. (2009). Nominal negative quantifiers as adjuncts. *Journal of Comparative Germanic Linguistics 12*, 5–30.

Bernini, G. and P. Ramat (1996). *Negative Sentences in the Languages of Europe*. Berlin: Mouton de Gruyter.

van Bergen, L. (2012). Ne + infinitive constructions in Old English. *English Language and Linguistics 16*(3), 487–518.

(2013). Let's talk about uton. In A. Jucker, D. Landert, A. Seiler, and N. Studer-Joho (Eds.), *Meaning in the History of English: Words and Texts in Context*, Studies in Language Companion Series, pp. 1–27. Amsterdam: John Benjamins.

Biber, D. (1995). *Dimensions of Register Variation: A Cross-Linguistic Comparison*. Cambridge: Cambridge University Press.

Biberauer, T. and I. Roberts (2011). Negative words and related expressions: a new perspective on some familiar problems. In P. Larrivée and R. Ingham (Eds.), *The Evolution of Negation*, pp. 23–60. Berlin: Mouton de Gruyter.

Birner, B. J. (2006). Semantic and pragmatic contributions to information status. *Acta Lingvistica Hafniensia 38*, 14–32.

Bobaljik, J. D. (1995). *Morphosyntax: the Syntax of Verbal Inflection*. Ph. D. thesis, Department of Linguistics and Philosophy, MIT.

Bobaljik, J. D. and H. Thrainsson (1998). Two heads aren't always better than one. *Syntax 1*, 37–71.

Borer, H. (1984). *Parametric Syntax: Case Studies in Semitic and Romance Languages*. Studies in Generative Grammar. Berlin: Mouton De Gruyter.

Breitbarth, A. (2009). A hybrid approach to Jespersen's cycle in West Germanic. *Journal of Comparative Germanic Linguistics 12*(2), 81–114.

Breitbarth, A. (2014). *The History of Low German Negation*. Oxford Studies in Diachronic and Historical Linguistics. Oxford: Oxford University Press.

Chomsky, N. (1995). *The Minimalist Program*. Cambridge, Mass.: MIT.

(1999). Derivation by phase. *MIT Occasional papers in linguistics 18*. 1–40

(2000). Minimalist inquiries: the framework. In R. Martin, D. Michaels, and J. Uriagereka (Eds.), *Step by Step: Essays on Minimalist Syntax in Honor of Howard Lasnik*, pp. 89–156. Cambridge, Mass.: MIT.

Dahl, O. (1979). Typology of sentence negation. *Linguistics 17*, 79–106.

Detges, U. and R. Waltereit (2002). Grammaticalisation vs. reanalysis: a semantic-pragmatic account of functional change in grammar. *Zeitschrift fur Sprachwissenschaft 21*(2), 151–195.

Dryer, M. S. (1996). Focus, pragmatic presupposition, and activated propositions. *Journal of Pragmatics 26*, 475–523.

Ecay, A. and M. Tamminga (2016). Persistence as a diagnostic of grammatical status: The case of Middle English negation. In R. Truswell and E. Mattieu (Eds.), *Proceedings of DIGS 15*. Oxford: Oxford University Press.

Espinal, M.-T. (1993). The interpretation of *no-pas* in catalan. *Journal of Pragmatics 19*, 353–369.

(2000). Expletive negation, negative concord and feature checking. *Catalan Working Papers in Linguistics 8*, 47–69.

Eythorsson, T. (2002). Negation in C: The syntax of negated verbs in Old Norse. *Nordic Journal of Linguistics 25*, 190–224.

Frisch, S. (1997). The change in negation in Middle English: A NegP licensing account. *Lingua 101*, 21–64.

Fulk, R. D. (2012). Anglian features in late West Saxon prose. In D. Denison, R. Bermúdez-Otero, C. McCully, and E. Moore (Eds.), *Analysing Older English*, pp. 63–74. Cambridge: Cambridge University Press.

Gałecki, A. and T. Burzykowski (2013). *Linear Mixed-Effects Models Using R: A Step-by-Step Approach*. Springer Texts in Statistics. New York: Springer.

Gardiner, A. H. (1904). The word iwn3. *Zeitschrift für Ägyptische Sprache und Altertumskunde 41*, 130–5.

van Gelderen, E. (2008). The Negative Cycle. *Linguistic Typology 12*, 195–243.

(2011). *The Linguistic Cycle*. Oxford: Oxford University Press.

Giannakidou, A. (2000). Negative...Concord? *Natural Language and Linguistic Theory 18*, 457–523.

(2006). N-words and negative concord. In K. v. H. C. Maienborn and P. Portner (Eds.), *Semantics: An International Handbook of Natural Language Meaning (Second edition)*. Berlin: Mouton de Gruyter.

Giorgi, A. (2004). From temporal anchoring to long distance anaphor binding. Paper presented at the 23rd West Coast Conference in Formal Linguistics. April 23–25, University of California at Davis, Davis CA.

Giorgi, A. and F. Pianesi (1997). *Tense and Aspect: From Semantics to Morphosyntax*. New York: Oxford University Press.

Gries, S. T. (2015a). The most under-used statistical method in corpus linguistics: multi-level (and mixed-effects) models. *Corpora 10*, 95–125.

(2015b). Some current quantitative problems in corpus linguistics and a sketch of some solutions. *Language and Linguistics 16*, 93–117.

Haeberli, E. (2001). Adjuncts and the syntax of subjects in Old and Middle English. In S. Pintzuk, G. Tsoulas, and A. Warner (Eds.), *Diachronic Syntax: Models and Mechanisms*. Oxford: Oxford University Press.

(2002a). Inflectional morphology and the loss of verb-second in English. In D. W. Lightfoot (Ed.), *Syntactic Effects of Morphological Change*, pp. 88–107. Oxford: Oxford University Press.

(2002b). Observations on the loss of verb-second in the history of English. In C. J. W. Zwart and W. Abraham (Eds.), *Studies in Comparative Germanic Syntax*. Amsterdam: John Benjamins.

(2011). Looking high and low for NegP in early English. In P. Larrivee and R. Ingham (Eds.), *The Evolution of Negation*, pp. 115–144. Berlin: Mouton de Gruyter.

Haeberli, E. and R. Ingham (2007). The position of negation and adverbs in Early Middle English. *Lingua 117*, 1–25.

Haegeman, L. (1995). *The Syntax of Negation*. Cambridge: Cambridge University Press.

(2001). Negative preposing, Negative inversion, and the split CP. In L. R. Horn and Y. Kato (Eds.), *Negation and Polarity: Syntactic and Semantic Perspectives*. Oxford: Oxford University Press.

Haegeman, L. and R. Zanuttini (1996). Negative concord in West Flemish. In A. Belletti and L. Rizzi (Eds.), *Parameters and Functional Heads: Essays in Comparative Syntax*. Oxford: Oxford University Press.

Hale, K. and S. J. Keyser (1993). On argument structure and the lexical expression of syntactic relations. In Kenneth Hale and Samuel J. Keyser (Eds.), *The View from Building 20: essays in honor of Sylvain Bromberger*. Cambridge Mass.: MIT Press.

Han, C.-H. (2001). The evolution of do-support in English imperatives. In S. Pintzuk, G. Tsoulas, and A. Warner (Eds.), *Diachronic Syntax: Models and Mechanisms*. Oxford: Oxford University Press.

Han, C.-H. and A. S. Kroch (2000). The rise of do-support in English imperatives: implications for clause structure. In *Proceedings of the 30th North East Linguistics Society*, pp. 311–325. GLSA.

Hansen, M.-B. M. (2009). The grammaticalisation of negative reinforcers in Old and Middle French. In M.-B. M. Hansen and J. Visconti (Eds.), *Current Trends in Diachronic Semantics and Pragmatics*, pp. 227–251. Bingley: Emerald Publishing.

Hansen, M.-B. M. and J. Visconti (2009). On the diachrony of "reinforced" negation in French and Italian. In C. R. C. Rossari and A. Spiridon (Eds.), *Grammaticalisation and Pragmatics: Facts, Approaches, Theoretical Issues*, pp. 137–171. Bingley: Emerald Publishing.

Hiraiwa, K. (2001). Multiple agree and the defective intervention constraint in Japanese. *MIT Working Papers in Linguistics 40*, 67–80.

Hoeksma, J. (2009). Jespersen recycled. In E. van Gelderen (Ed.), *Cyclical Change*, pp. 15–35. Amsterdam: John Benjamins.
(2012). On the natural history of negative polarity items. *Linguistic Analysis 38*, 3–33.
(2014). The Middle Dutch negative clitic: Status, position and disappearance. *Lingua 147*, 50–68.
Ingham, R. (2005). The loss of Neg-V-to-C in Middle English. *Linguistische Berichte 202*, 171–206.
(2006). On two types of negative concord in early English. *Language Variation and Change 18*(3), 241–266.
(2007). NegP and negated constituent movement in the history of English. *Transactions of the Philological Society 105*, 365–397.
(2011). Ne-drop and indefinites in Anglo-Norman and Middle English. In P. Larrivée and R. Ingham (Eds.), *The Evolution of Negation*, pp. 145–163. Berlin: Mouton de Gruyter.
Israel, M. (1998). *The Rhetoric of Grammar: Scalar Reasoning and Polarity Sensitivity*. Ph. D. thesis, University of California.
Jack, G. B. (1978a). Negation in Later Middle English prose. *Archivum Linguisticum 9*, 58–72.
(1978b). Negative adverbs in Early Middle English. *English Studies 59*, 295–309.
Jaeger, A. (2008). *A History of German Negation*. Amsterdam: John Benjamins.
Jespersen, O. (1917). *Negation in English and other Languages*. Kobenhavn: Ejnar Munksgaard.
Jespersen, O. (1924). *The Philosophy of Grammar*. London: Allen and Unwin.
Johnson, D. E. (2008). Getting of the Goldvarb standard: introducing Rbrul for mixed-effects variable rule analysis. *Language and Linguistics Compass 3*, 359–383.
van Kemenade, A. (1999). Sentential negation and clause structure in Old English. In I. T.-B. van Ostade, G. Tottie, and W. van der Wurff (Eds.), *Negation in the history of English*. Berlin: Mouton de Gruyter.
(2001). Jespersen's cycle revisited: formal properties of grammaticalisation. In S. Pintzuk, G. Tsoulas, and A. Warner (Eds.), *Diachronic Syntax: models and mechanisms*. Oxford: Oxford University Press.
(2011). Secondary negation and information structure organisation in the history of English. In P. Larrivee and R. Ingham (Eds.), *The Evolution of Negation*, pp. 77–114. Berlin: Mouton de Gruyter.
Kiparsky, P. and C. Condoravdi (2006). Tracking Jespersen's cycle. In B. J. M. Janse and A. Ralli (Eds.), *Proceedings of the 2nd International Conference of Modern Greek Dialects and Linguistic Theory*. Mytilene. Doukas.
Klima, E. S. (1964). Negation in English. In J. Fodor and J. Katz (Eds.), *The Structure of Language*. Eaglewood Cliffs, New Jersey: Prentice-Hall.
Koopman, W. (2005). Transitional syntax: postverbal pronouns and particles in Old English. *English Language and Linguistics 9*, 47–62.
Kroch, A., B. Santorini, and A. Diertani (Eds.) (2004). *Penn-Helsinki Parsed Corpus of Early Modern English*. Philadelphia: Department of Linguistics, University of Pennsylvania.
Kroch, A. S. (1989). Reflexes of grammar in patterns of language change. *Language Variation and Change 1*, 199–244.

(1994). Morphosyntactic variation. In K. Beals (Ed.), *Proceedings of the Thirtieth Annual Meeting of the Chicago Linguistics Society*, Chicago, pp. 180–201. Chicago Linguistics Society.

Kroch, A. S. and A. Taylor (1997). Verb movement in Old and Middle English: dialect variation and language contact. In A. van Kemenade and N. Vincent (Eds.), *Parameters of Morphosyntactic Change*. Oxford: Oxford University Press.

(2000). *Penn-Helsinki Parsed Corpus of Middle English* (Second ed.). Philadelphia: Department of Linguistics, University of Pennsylvania.

(2000). Verb-object order in Early Middle English. In S. Pintzuk, G. Tsoulas, and A. Warner (Eds.), *Diachronic Syntax: Models and Mechanisms*. Oxford: Oxford University Press.

Labov, W. (1972). Negative attraction and negative concord in English grammar. *Language 48*, 773–818.

Labrum, R. W. (1982). *Conditions on Double Negation in the History of English with Comparison to Similar Developments in German*. Ph. D. thesis, Stanford University, Stanford.

Ladusaw, W. (1992). Expressing negation. In C. Barker and D. Dowty (Eds.), *Proceedings of the Second Conference on Semantics and Linguistic Theory*, Ohio State University Working Papers in Linguistics, vol. 40, pp. 237–259. Columbus: The Ohio State University.

(1993). Negation, indefinites and the Jespersen cycle. In *Proceedings of the Nineteenth Annual Meeting of the Berkeley Linguistics Society, February 12-15, 1993*. Berkeley Linguistic Society.

Larrivée, P. (2015). The markedness of double negation. In *Negation and Polarity: Experimental Perspectives*, pp. 177–198. New York: Springer.

Levin, B. and M. Hovav (1995). *Unaccusativity: At the Syntax-lexical Semantics Interface*. Linguistic inquiry monographs. Cambridge Mass.: MIT Press.

Lewis, R. E. (Ed.) (2001). *Middle English Dictionary*. Ann Arbor: University of Michigan.

Merchant, J. (2013). Polarity items under ellipsis. In L. Cheng and N. Corver (Eds.), *Diagnosing Syntax*. Oxord: Oxford University Press.

Mitchell, B. (1985). *Old English Syntax*. Oxford: Oxford University Press.

Morris, R. (Ed.) (1873). *Old English Homilies of the Twelfth Century. Second series.*

Nash, L. and A. Rouveret (1997). Proxy categories in phrase structure theory. In K. Kusumoto (Ed.), *Proceedings of NELS 27*, pp. 287–304. Amherst: GLSA.

Nevalainen, T. (1996). Social mobility and the decline of multiple negation in Early Modern English. In J. Fisiak and M. Krygier (Eds.), *Advances in English Historical Linguistics*. Berlin: Mouton de Gruyter.

(1997). Recycling inversion: The case of initial adverbs and negators in Early Modern English. *Studia Anglica Posnaniensia 31*, 203–214.

Nevalainen, T. and H. Ramoulin-Brunberg (2003). *Historical Sociolinguistics: language change in Tudor and Stuart England*. London: Longman.

Nunes, J. (1995). The diachronic distribution of bare and prepositional infinitives in English. In H. Andersen (Ed.), *Historical Linguistics 1993*. Amsterdam: John Benjamins.

Pesetsky, D. and E. Torrego (2007). The syntax of valuation and the interpretability of features. In *Phrasal and Clausal Architecture: Syntactic Derivation and Interpretation*, pp. 262–294. Amsterdam: John Benajmins.

Pintzuk, S. (1999). *Phrase Structures in Competition: Variation and Change in Old English Word Order*. New York: Garland.

(2000). Verb-object order in Old English: variation as grammatical competition. In David Lightfoot (ed.), *Syntactic Effects of Morphological Change*, pp. 276–299. Oxford: Oxford University Press.

(2005). Arguments against a universal base: evidence from Old English. *English Language and Linguistics 9*, 115–138.

Pintzuk, S. and A. Taylor (2003). The loss of OV order in the history of English. In A. van Kemenade and B. Los (Eds.), *Blackwell Handbook of the History of English*. Oxford: Blackwell.

Preminger, O. (2013). That's not how you agree: A reply to Zeijlstra. *The Linguistic Review 30*, 491–500.

Rauer, C. (Ed.) (2013). *The Old English Martyrology: Edition, Translation and Commentary*. Woodbridge and Rochester: Boydell and Brewer.

Rissanen, M. (1994). The position of not in Early Modern English questions. In D. Kastovsky (Ed.), *Studies in Early Modern English*. Berlin: Mouton de Gruyter.

(1999). Isn't it? or is it not? on the order of postverbal subject and negative particle in the history of english. In I. T.-B. van Ostade, G. Tottie, and W. van der Wurff (Eds.), *Negation in the History of English*. Berlin: Mouton de Gruyter.

Roberts, I. (1993). *Verbs and Diachronic Syntax: A comparative History of English and French*. Dordrecht: Kluwer.

(2010). Head movement and the Minimalist Program. In C. Boeckx (Ed.), *The Oxford Handbook of Syntactic Minimalism*. Oxford: Oxford University Press.

Roberts, I. and A. Roussou (2003). *Syntactic Change: A Minimalist Approach to Grammaticalisation*. Cambridge: Cambridge University Press.

Rowlett, P. (1997). Jespersen, negative concord and A'-binding. In D. Forget, P. Hirschbuhler, F. Martineau, and M.-L. Rivero (Eds.), *Negation and polarity: syntax and semantics. Selected papers from the colloquium negation: syntax and semantics, Ottowa 11-13 May 1995*. Amsterdam: John Benjamins.

(1998). *Sentential Negation in French*. Oxford: Oxford University Press.

Rowley, S. (2011). *The Old English Version of Bede's Historia Ecclesiastica*. Woodbridge and Rochester: Boydell and Brewer.

Sankoff, D. and W. Labov (1979). On the uses of variable rules. *Language in Society 9*, 189–222.

Schwegler, A. (1988). Word-order changes in predicate negation strategies in Romance languages. *Diachronica 5*, 21–58.

Schwenter, S. (2006). Fine tuning Jespersen's cycle. In B. J. B. Laurence R. Horn and G. L. Ward (Eds.), *Drawing the Boundaries of Meaning: Neo-Gricean Studies in Pragmatics and Semantics in Honor of Laurence R. Horn*, pp. 327–344. Amsterdam: John Benjamins.

Smith, J. (2001). Negative concord in the old and new world: evidence from Scotland. *Language Variation and Change 13*, 109–134.

Sweet, H. (Ed.) (1958 (1871)). *King Alfred's West-Saxon Version of Gregory's Pastoral Care*. EETS 45, 50. Oxford: Oxford University Press.

Tagliamonte, S. A. (2006). *Analysing Sociolinguistic Variation*. Key Topics in Sociolinguistics. Cambridge: Cambridge University Press.

Tagliamonte, S. A. and R. Baayen (2012). Models, forests and trees of York English: Was/were variation as a case study for statistical practice. *Language Variation and Change 24*, 135–178.

Taylor, A., A. Nurmi, A. Warner, S. Pintzuk, and T. Nevalainen (Eds.) (2006). *Parsed Corpus of Early English Correspondence*. Compiled by the CEEC Project Team. York: University of York and Helsinki: University of Helsinki. Distributed through the Oxford Text Archive.

Taylor, A. and S. Pintzuk (2012). Verb order, object position and information status in Old English. *York Papers in Linguistics 2 12a*, 29–52.

Taylor, A., A. Warner, S. Pintzuk, and F. Beths (Eds.) (2002). *The York-Toronto-Helsinki Parsed Corpus of Old English Prose*. York: Department of Language and Linguistic Science, University of York.

Tottie, G. (1991a). Lexical diffusion in syntactic change: Frequency as a determinant of linguistic conservatism in the development of negation in English. In D. Kastovsky (Ed.), *Historical English Syntax*. Berlin: Mouton de Gruyter.

(1991b). *Negation in English Speech and Writing: A Study in Variation*. Quantitative Analyses of Linguistic Structure. San Diego: Academic Press.

Ukaji, M. (1999). On the scope of negative concord. In I. T.-B. van Ostade, G. Tottie, and W. van der Wurff (Eds.), *Negation in the History of English*. Berlin: Mouton de Gruyter.

van der Auwera, J. (2009). The Jespersen cycles. In E. van Gelderen (Ed.), *Cyclical Change*, pp. 35–71. Amsterdam: John Benjamins.

van der Auwera, J. and L. Alsenoy (2011). Indefinite pronouns, synchrony and diachrony: comments on Willis. In P. Larrivée and R. Ingham (Eds.), *The Evolution of Negation*, pp. 325–346. Berlin: Mouton de Gruyter.

Waite, G. (2013). The Old English Bede and the glosses in the Tiberius Bede. *Parergon 30*, 1–49.

Wallage, P. W. (2005). *Negation in Early English: parametric variation and grammatical competition*. Ph. D. thesis, Department of Language and Linguistic Science, University of York.

(2008). Jespersen's cycle in Middle English: parametric variation and grammatical competition. *Lingua 118*, 643–674.

(2012a). Negative inversion, negative concord and sentential negation in the history of english. *English Language and Linguistics 16*(1), 3–33.

(2012b). Quantitative evidence for a feature-based account of grammaticalization in English: Jespersen's cycle. In E. Traugott and T. Nevalainen (Eds.), *The Oxford Handbook of the History of English*. Oxford: Oxford University Press.

(2013). Functional differentiation and grammatical competition in the English Jespersen cycle. *Journal of Historical Syntax 2*(1).

Warner, A. R. (1993). *English Auxiliaries: Structure and History*. Cambridge: Cambridge University Press.

(2005). Why DO dove: evidence for register variation in Early Modern English negatives. *Language Variation and Change 17*, 257–280.

Whitelock, D. (1962). *The Old English Bede*. British Academy. Sir Israel Gollancz memorial lecture. Oxford: Oxford University Press.

Willis, D. (2011a). A minimalist approach to Jespersen's cycle in Welsh. In D. Jonas, J. Whitman, and A. Garrett (Eds.), *Grammatical Change: Origins, Natures, Outcomes*, pp. 93–119. Oxford: Oxford University Press.

(2011b). Negative polarity and the quantifier cycle: comparative diachronic perspectives from european languages. In P. Larrivée and R. Ingham (Eds.), *The Evolution of Negation*, pp. 285–324. Berlin: Mouton de Gruyter.

Willis, D., C. Lucas, and A. Breitbarth (Eds.) (2013). *Negation in the Languages of Europe and the Mediterranean: Volume 1 Case Studies*. Oxford: Oxford University Press.

van der Wouden, T. (1994). *Negative Contexts*. Ph. D. thesis, Department of Dutch, University of Groningen.

van der Wurff, W. (1997). Deriving object-verb order in late Middle English. *Journal of Linguistics 33*, 485–509.

(1999a). Objects and verbs in modern Icelandic and fifteenth century English: a word order parallel and its causes. *Lingua 109*, 237–265.

(1999b). On expletive negation with adversative predicates in the history of English. In I. T.-B. van Ostade, G. Tottie, and W. van der Wurff (Eds.), *Negation in the History of English*. Berlin: Mouton de Gruyter.

Wurmbrand, S. (2012). The syntax of valuation in auxiliary–participle constructions. In J. Choi, E. A. Hogue, J. Punske, D. Tat, J. Schertz, and A. Trueman (Eds.), *Coyote Working Papers: Proceedings of the 29th West Coast Conference on Formal Linguistics (WCCFL29)*. Somerville Mass.: Cascadilla Press.

Yoon, S. (2011). *Not in the Mood: the Syntax, Semantics and Pragmatics of Evaluative Negation*. PhD, University of Chicago, Chicago.

Zeijlstra, H. H. (2004). *Sentential Negation and Negative Concord*. Ph. D. thesis, Universiteit van Amsterdam.

Zeijlstra, H. H. (2008). Negative concord is syntactic agreement. Available at: Lingbuzz/000645.

(2010). On French negation. In *Proceedings of the 35th annual meeting of the Berkely Linguistics Society*, pp. 447–458. Berkley CA.

(2012). There is only one way to agree. *The Linguistic Review 29*, 491–530.

Zwicky, A. and G. K. Pullum (1983). Cliticization vs. Inflection: English *n't*. *Language 59*, 502–513.

# Index of names

# Index of subjects